The Basics of Engineering

THE BASICS OF ENGINEERING

Lokesh Pandey

ARCLER
PRESS

www.arclerpress.com

The Basics of Engineering

Lokesh Pandey

Arcler Press
224 Shoreacres Road
Burlington, ON L7L 2H2
Canada
www.arclerpress.com
Email: orders@arclereducation.com

ISBN: 978-1-77469-472-5 (Hardcover)

ABOUT THE AUTHOR

Lokesh Pandey is currently pursuing his PhD in Mechanical Engineeing from Uttarakhand Technical University, India, where he also completed his M.Tech in Thermal Engineeing. He has more than 3 years teaching experience as a Faculty for Mechanical Engineering. He has published articles in reputed Journals and has also chaired many conferences and workshops.

TABLE OF CONTENTS

LIST OF FIGURES

GLOSSARY

A

Additive Manufacturing – Additive manufacturing is the process of creating an object by building it one layer at a time. It is the opposite of subtractive manufacturing, in which an object is created by cutting away at a solid block of material until the final product is complete.

Architects – A person who designs buildings and in many cases also supervises their construction.

B

Binomial Distribution – A frequency distribution of the possible number of successful outcomes in a given number of trials in each of which there is the same probability of success.

Biomedical Engineering – Biomedical engineering is the application of the principles and problem-solving techniques of engineering to biology and medicine.

C

Canals – A canal is a human-made waterway that allows boats and ships to pass from one body of water to another.

Cathode Ray Tube – Vacuum tube that produces images when its phosphorescent surface is struck by electron beams. CRTs can be monochrome (using one electron gun) or color (typically using three electron guns to produce red, green, and blue images that, when combined, render a multicolor image).

Civilization – A civilization (or civilization) is any complex society characterized by the development of a political state, social stratification, urbanization, and symbolic systems of communication beyond natural spoken language (namely, a writing system).

Computer-Aided Design – Computer-aided design is a way to digitally create 2D drawings and 3D models of real-world products—before they're ever manufactured. With 3D CAD, you can share, review, simulate, and modify designs easily, opening doors to innovative and differentiated products that get to market fast.

Conductivity – Conductivity is the measure of the ease at which an electric charge or heat can pass through a material. A conductor is a material which gives very little resistance to the flow of an electric current or thermal energy. Materials are classified as metals, semiconductors, and insulators.

Corrosion Resistance – Corrosion resistance can be defined as the ability to protect the substrate from corrosion. In this case coating microstructure, in particular the appearance of open porosity and cracks, can be more important than the coating composition.

Creep Resistance – Creep resistance is a term used in materials science that refers to a solid material's ability to resist "creep," which refers to the tendency of a material to slowly deform over a long period of exposure to high levels of stress.

D

Dams – A dam is a structure built across a river or stream to hold back water. People have used different materials to build dams over the centuries.

Ductility – Ductility is a mechanical property commonly described as a material's amenability to drawing. In materials science, ductility is defined by the degree to which a material can sustain plastic deformation under tensile stress before failure.

E

Engineering – Engineering is the designing, testing, and building of machines, structures, and processes using math and science. Studying it can lead to a rewarding career. Engineering is a discipline dedicated to problem solving.

Entrepreneurship – The activity of setting up a business or businesses, taking on financial risks in the hope of profit.

Evolution – Evolution is a process that results in changes in the genetic content of a population over time.

H

Heat Resistance – Heat resistant materials are materials that can protect various elements from heat generated due to high temperature operations. They can reduce the chances of dangerous off-gassing hazards because of heating of sensitive parts like wires, cables, and refrigeration lines.

Histogram – A diagram consisting of rectangles whose area is proportional to the frequency of a variable and whose width is equal to the class interval.

I

Interpersonal Skills – Interpersonal skills are the behaviors and tactics a person uses to interact with others effectively. In the business world, the term refers to an employee's ability to work well with others. Interpersonal skills range from communication and listening to attitude and deportment.

Irrigation – The supply of water to land or crops to help growth, typically by means of channels.

M

Magnetic Resonance Imaging (MRI) – Magnetic resonance imaging (MRI) is a medical imaging technique that uses a magnetic field and computer-generated radio waves to create detailed images of the organs and tissues in your body. Most MRI machines are large, tube-shaped magnets.

Mechanical Engineering – Mechanical engineering is the study, design, development, construction, and testing of mechanical and thermal sensors and devices, including tools, engines, and machines.

N

Nuclear Energy – Nuclear energy is the energy in the nucleus, or core, of an atom. Atoms are tiny units that make up all matter in the universe, and energy is what holds the nucleus together. There is a huge amount of energy in an atom's dense nucleus.

P

Philosophy – The study of the fundamental nature of knowledge, reality, and existence, especially when considered as an academic discipline.

Poisson Distribution – A discrete frequency distribution which gives the probability of a number of independent events occurring in a fixed time.

Probability – The extent to which an event is likely to occur, measured by the ratio of the favorable cases to the whole number of cases possible.

Prosthetic – An artificial feature or piece of flexible material applied to a person's face or body to change their appearance temporarily.

Prototyping – A prototype is a draft version of a product that allows you to explore your ideas and show the intention behind a feature or the overall design concept to users before investing time and money into development.

R

Robotics – The branch of technology that deals with the design, construction, operation, and application of robots.

S

Sensor – A sensor is a device that detects and responds to some type of input from the physical environment. The specific input could be light, heat, motion, moisture, pressure, or any one of a great number of other environmental phenomena.

Spatial Distribution – A spatial distribution in statistics is the arrangement of a phenomenon across the Earth's surface and a graphical display of such an arrangement is an important tool in geographical and environmental statistics.

Statistics – The practice or science of collecting and analyzing numerical data in large quantities, especially for the purpose of inferring proportions in a whole from those in a representative sample.

T

Thermal Conductivity – Thermal conductivity can be defined as the rate at which heat is transferred by conduction through a unit cross-section area of a material, when a temperature gradient exits perpendicular to the area.

V

Valley – A low area of land between hills or mountains, typically with a river or stream flowing through it.

LIST OF ABBREVIATIONS

A.D.	Anno Domini
ABET	Accreditation Board for Engineering and Technology
ACIS	American Committee for Interoperable Systems
AI	artificial intelligence
AM	additive manufacturing
B.C.	Before Christ
CAD	computer-aided design
CEOs	Chief Executive Officers
CID	communication across disciplines
CPD	collaborative product design
CPUs	computer processing units
CRT	cathode ray tube
CS	computer science
CSE	computer science engineering
DC	direct current
DFA	design for assembly
DFM	design for manufacturing
DSSs	decision support systems
DVD	digital versatile disc
DXF	drawing exchange format
EDM	engineering data management
EE	electrical engineering
EMI	electromagnetic interference
ERP	enterprise resource planning
FE	fundamentals of engineering exam
GIS	geographic information system
GPS	global positioning system
HMDs	helmet-mounted displays

ICG	interactive computer graphics
ICs	integrated circuits
IGES	initial graphics exchange specification
LCDs	liquid crystal displays
MBA	master of business administration
MEMS	microelectromechanical systems
MPI	Metal Processing Institute
MRI	magnetic resonance imaging
NASA	National Aeronautics and Space Administration
NSPE	National Society of Professional Engineers
PDM	product data management
PE	professional engineer
PIM	product information management
QDs	quantum dots
SaaS	software as a service
SAGE	semi-automatic ground environment
STEP	Standard for the Exchange of Products
TDM	technical document management
TIM	technical information management
TM	trade mark
TV	television
TVA	Tennessee Valley Authority's
UMC	unit manufacturing cost
US	United States
VR/AR	virtual and augmented reality
VRML	virtual reality modeling language
WWII	World War II
XML	extensible mark-up language

PREFACE

This book introduces the readers to the basics of engineering. This book sheds light on engineering as a profession, basics of mechanical engineering, engineering communication and ethics, engineering materials and their applications, mathematics, probability, and statistics in engineering and applications of engineering in different fields.

The first chapter stresses on the basic significance of engineering and its history, fields of specialization, introduction to the engineering profession, etc. This chapter will also emphasize on the introduction to engineering technology and engineering design process.

The second chapter takes the readers through an overview of engineering as a profession. This chapter will provide highlights on what is engineering and what engineers do, some areas of engineering, preparing for an engineering career, engineering profession and a 21st-century engineer. The chapter also explains societal issues and engineering, redefining the ever-evolving engineering profession, the decreased durability of initial engineering education and determining whether engineering is a profession here or there.

Then, the third chapter explains the basics of mechanical engineering. It also explains the basic concepts, CAD design, CAD software, 2D graphics software, graphical representation of image data, and analysis software. This chapter also sheds light on CAD standards and translators, applications of CAD and product design for manufacturing (DFM) and assembly.

The fourth chapter introduces the readers to the engineering communication and ethics. This chapter explains the importance of ethics and communication in engineering, professional code of ethics, ethical theories and the importance of ethical conduct in engineering. The chapter also addresses different parts of a communication system, types of signals and some moral issues in engineering.

The fifth chapter throws light on Engineering materials and their applications. This chapter contains the importance of engineering materials in the modern world, evolution of engineering materials, current trends and advances in materials, classification, application of engineering materials, etc. This chapter also sheds light on the future engineering materials, materials and the environment green design and gives an introduction to material selection.

The sixth chapter takes the readers through the concept of mathematics, probability, and statistics in engineering. The readers are then told about the organization of text, probability tables, and computer software, probability, and random variables, probability distribution, some important discrete distribution, important continuous distribution, etc. It also explains observed data and a graphical representation, and gives an introduction to statistics, descriptive statistics, enumerative versus analytical statistics, and collection of data.

The last chapter of this book sheds light on the application of engineering in various fields. This chapter also mentions the different types of engineering, like civil, mechanical, electrical, computer science, IT engineering, etc., and the applications of each in different fields.

This book has been designed to suit the knowledge and pursuit of the researcher and scholars and to empower them with various aspects of engineering and technology on which the modern world is based, so that they are updated with the information. I hope that the readers find the book explanatory and insightful and that this book is referred by the scholars across various fields.

CHAPTER 1

INTRODUCTION TO ENGINEERING

CONTENTS

The term engineering is derived from the Latin words "Ingenium," which means "cleverness," and "ingeniare," which means "to devise/concoct." Engineering is the use of scientific, economic, social, and practical knowledge to create, design, build, maintain, and improve structures, machines, devices, systems, materials, and strategies.

The topic of engineering is quite broad and includes a variety of more specialized professions of engineering, each with a more particular concentration on certain areas of implemented technology, generating, and utility patterns. Engineers play an important role in defining and progressing our society by providing tailored solutions to specific situations through integrated knowledge of engineering concepts and basic sciences.

Engineers can create solutions by combining knowledge from several sources, logic, and computation. Today's engineers are entirely prepared with themes and programs that must be developed to meet the impending worldwide demands.

1.1. HISTORY OF ENGINEERING

Engineering did not come into being by royal edict or legislative fiat. Over more than 50 centuries of documented history, it has evolved and developed as a practical craft and a vocation. Its origins may be traced back to the start of civilization, and its progression tracks that of humanity.

Our forefathers strove to manage and harness natural resources and forces for public advantage, just as we do now. They researched and observed natural laws, gaining knowledge of mathematics and science that the common people lacked.

They used this information with discretion and judgment to meet good societal requirements with ports, roads, buildings, irrigation, and flood control systems, as well as other creative works in engineering history, teach us to respect the unknown and its accomplishments (Figure 1.1).

Figure 1.1. Image showing an engineer.

Source: Image by pixabay.

They enable us to see the present in the context of the past, to identify characteristics, and compare the causes of the great changes that have marked human history. We may feel the broad glide of history and look at the presentation as a component of that drift by investigating the beginnings of engineering.

This allows us to better understand our wants, objectives, and behavior by placing them in their proper perspective. The goal of this chapter is to provide a concise overview of the evolution of engineering from the earliest documented times to the present.

This, of course, is a dircctive. This is, of course, a major project, and with the limited space available, we will only quickly describe the highlights of engineering records.

1.1.1. Engineering in the Early Civilizations: The Mesopotamians

The ancient inhabitants of Mesopotamia, the territory between the Tigris and Euphrates rivers that is now Iraq, must be attributed with significant technical achievements. The wheeled cart is thought to have originally arrived in this area.

At the dawn of recorded history, the ancient and enigmatic Sumerian people-built canals, temples, and city walls that formed the world's first engineering achievements in southern Mesopotamia. Mesopotamia's terrain was vulnerable to invasion from the north, east, and west, and its history is a jumbled chronicle of conquests and occupations by surrounding peoples.

The Babylonians and Assyrians were the most powerful monarchs of ancient. Records engraved on clay tablets have been uncovered and interpreted, revealing insight into life thousands of years ago in that region. These documents reveal that an angle measuring instrument known as the astrolabe was used for astronomical observations as early as 2000 B.C.

This device, which consisted of a graded circle and a sighting arm, was based on the Mesopotamian 60-unit numeral system. That system has survived to the current day in time and angle measurements. The ziggurat, a temple tower created in honor of their gods, was the most peculiar type of construction left by the Mesopotamians.

The ziggurat was a brick tiered pyramid with staircases and setbacks, as well as a shrine or chapel at the top. This style of construction is said to have been the Tower of Babel referenced in the Old Testament. Hammurabi, Babylonia's great monarch who reigned for 43 years (approximately 1850 to 1750 B.C.), developed a completely new system of law that carries his name.

This historic regulation, which is considered a predecessor of today's building standards, imposed fines on anyone who condoned improper construction techniques. The Hammurabi Code sent a vital message about quality assurance and professional accountability, and it imposed extraordinarily harsh penalties for violations.

It is hardly unexpected that the people who lived in the Tigris and Euphrates valleys created major irrigation and flood control systems. In Iraq, evidence of abandoned canals may still be found in the form of embankments, lakes, and streams. The Nahrwan, a 400-foot-wide canal that ran mainly parallel to the Tigris River for 200 miles, irrigated an area that averaged 18 miles in width.

Mesopotamians employed imposing brick dams to channel tiny torrents into the canal. During King Sennacherib's reign, the Assyrians built the earliest prominent example of a public water system (about 700 B.C.). They constructed a 30-mile-long feeder canal to carry fresh water from the slopes of Mount Tas to the existing Khosr reservoir River, through which the water flowed further 15 kilometers into Ninevah.

An elevated cut-stone aqueduct was erected at Jerwan to carry the open canal over a small brook. This renowned construction measured 863 feet long, 68 feet broad, and 28 feet tall at its peak. It supported a canal that was about 50 feet broad and 5 feet deep. The canal was supported by a thick layer of concrete, the first recorded application of this building material.

1.1.2. Engineering in the Early Civilizations: The Egyptians

Experts in planning and building evolved during the ancient Egyptian civilization. These engineering forefathers held high-ranking positions as valued advisors to the Egyptian monarchs. This job was filled by a general building specialist known as the king's "head of works."

These ancient engineers/architects conducted the first known type of surveying, devised efficient irrigation systems, and erected magnificent stone structures. The periodic flooding of the Nile necessitated the re-establishment of land borders. Egyptian engineers carried out these inspections using pieces of rope that had been soaked in water, dried, and then coated with a wax compound to ensure continuous length.

They may have employed rudimentary surveying devices as well, although none have been discovered. The Egyptians created and maintained an enormous system of dykes, canals, and drainage systems as early as 3300 B.C. A vast population inhabited the Nile's narrow fertile valley, and irrigation works were required to sustain the enormous population and utilize the skill of agriculture.

Because horses, wheeled vehicles, and roads were not available in Egypt until before 1785 B.C., the river functioned as the primary mode of transportation. Ancient Egyptian engineers aimed to construct the world's tallest, widest, and most lasting buildings. Their palaces, temples, and tombs were built to represent triumphant and eternal power.

The pyramids are the most well-known creations of Egyptian architects at Sakkara. Imhotep erected the Step Pyramid in Sakkara in around 2980 B.C. as a burial site for the monarch Zoser.

The ancient Egyptians considered the king's tomb to be a residence where he truly lived after death, and some of the most sophisticated mastabas had multiple chambers and storage cells where food and weaponry were kept near to the deceased monarch and his household. Zoser's Step Pyramid consists of six mastabas, with the second built on top of the first, the third on top of the second, and so on (Figure 1.2).

Figure 1.2. Image showing Egyptian sculpture.

Source: Image by pixabay.

Imhotep may have erected the mastabas one on top of the other to prevent tomb thieves from excavating into the building from the top, as was typical. Three pyramids still exist on the west bank of the Nile River in Giza as memorials of the Egyptians' exceptional engineering talents.

The greatest pyramid, known as the Great Pyramid or the Pyramid of Cheops, stands 481 feet tall and has a 13-acre base. The pyramid is made up of more than 2 million stone pieces, each weighing 2.5 tons. Some of the internal blocks are as heavy as 30 tons.

1.1.3. Contributions of the Greeks

Beginning around 600 B.C., the Greek style of life and ideas started to dominate the eastern Mediterranean region. The Greeks are primarily recognized for their abstract reasoning and ability to conceptualize and integrate historical information. Their achievements in art, literature, and philosophy overshadowed their contributions to engineering.

They tended to priorities theory above testing and verification, as well as practical application. Indeed, the great Greek thinkers held that any application of the results of the mind to worldly demands was deserving of neither dignity nor respect. Nonetheless, the Greek architect on made the first significant step toward professional standing.

He was renowned as a master builder and building specialist with knowledge and skills that were beyond the common citizen's capabilities. Because the Greek peninsula was so divided by mountain ranges, ground transportation was impossible. The Greeks became the first major port builders by turning to the sea. Herodotus describes a massive breakwater or mole built to safeguard Samos' harbor.

The breakwater was 400 yards long and erected at 120 feet of water. This was the earliest known construction of an artificial harbor, and it served as a model for harbor planning even until contemporary times. The Greeks' interest in navigation led to the development of the construction of the world's first lighthouse, the Pharos, in the port of Alexandria.

This 370-foot-tall monument, which was erected around 300 B.C., was considered one of the Seven Wonders of the Ancient World. Under the leadership of the architecton Eupalinus of Megara, a 3300-foot-long tunnel was constructed through a 900-foot slope on the island of Samos. The main tunnel was roughly 5.5 feet wide and tall, and was hand-chiseled through solid limestone.

A ditch 30 feet deep and 3 feet broad was dug at the bottom of the main tunnel. Water was transported to the city using clay pipes in this trench. The tunnel was built from both ends; however, the surveying procedures were different the procedures employed to do this task remain unknown.

During Greece's Golden Age, the king Pericles embarked on a massive construction project to create Athens the most beautiful city on the planet. He hired famous artists and builders of the day to construct temples, shrines, and sculptures atop the Acropolis, the flat-topped rock overlooking the city. The remains of these constructions are now one of the world's most spectacular sites.

1.1.4. Contributions of the Romans

The Romans, the most famous engineers of antiquity, committed more of their resources to public works than their predecessors. They constructed arenas, roads, aqueducts, temples, town halls, baths, and public forums with inexpensive labor, including thousands of slaves, and abundant raw materials.

Scholars split Roman history into two major periods: (1) the Republic, which lasted from the supposed date of Rome's birth in 535 B.C. until 24 B.C.; and (2) the Empire, which lasted from 24 B.C. until A.D. 476. The Republic was a period of conquest and exploitation of Rome's vast colonial domains, during which Roman engineering achievements were mostly limited to Italy.

During the Empire, public works were expanded into the colonies; remnants of some of these engineering projects may still be found today in Spain, France, North Africa, and the Near East. The Romans, unlike the Greeks, were practical builders who relied on experience rather than mathematical reasoning and science. Their creations were basic in concept but grand in scale and execution.

In general, their creations stressed usefulness over creative or esthetic value. Roman builders are credited with substantial contributions to engineering, including better building processes, the discovery and use of hydraulic cement, and the development of a variety of construction machinery such as pile drives, treadmill hoists, and wooden bucket wheels.

1.1.5. Engineering in the Middle Ages

There were minimal improvements in engineering during the roughly eight centuries after the collapse of the Roman Empire, known as the Middle Ages. During this time, there was modest technical advancement, particularly in structural design and the creation of energy-saving and power-enhancing machinery and gadgets.

The Gothic cathedrals were among the most intriguing constructions of the Middle Ages. These buildings have been described as "among the lightest, most daring 'skeleton stone' constructions ever attempted by man." These towering and beautiful constructions have stained glass windows, pointed central arches, and high thin walls supported around the sides by half arches known as flying buttresses.

The engineer/architects who planned them and the skilled craftsmen who built them both demonstrated a high level of structural ability. Massive fortress residences or castles were erected at an era when powerful landowners sought to secure themselves and their lands. Thick walls, lofty defensive towers, and a surrounding broad ditch bridged by a single bridge distinguished the fortress home.

The development of gunpowder and cannons (approximately A.D. 1500) put an end to the construction of medieval castles. Engineers strove to augment or complement the productive capacities of people and animals during the Middle Ages by creating and refining labor-saving devices.

During this time, the windmill was invented, while water mills were refined and employed in novel ways. By A.D. 700, water wheels for mill propulsion were in use throughout Europe. The spinning wheel and a hinged rudder for ships were two further mechanical improvements that arose in Europe during the Middle Ages. By A.D. 900, the Vikings had mastered shipbuilding and explored Greenland.

The term engineer was initially used in the Middle Ages (about A.D. 1000–1200). The terms "engine" and "ingenious" derive from the Latin phrase "in generare," which means "to produce." Thus, the person who invented or developed war machines like as battering rams, catapults, and assault towers became known as the inventor or "engine-er."

1.1.6. Advancements in Engineering: A.D. 1750–1900

Mining, industry, and transportation all advanced throughout the 150 years before the twentieth century. James Watt conceived and built a functioning model of a greatly improved steam engine in the 1760s. He produced hundreds of the engines with the help of maker Matthew Boulton.

By 1800, there were 500 Boulton and Watt engines in operation in Britain, pumping out mines and powering machinery in iron industries and textile mills. Until the mid-1700s, charcoal was utilized for iron ore smelting Because of a lack of wood for charcoal production, iron makers began employing coke, a lighter, more porous type of coal, for the smelting process.

The rising need for coke necessitated the construction of steam-driven mine pumps to rid the coal mines of water. The new source of electricity was quickly put to use in the ironworks, moving machinery and operating new blowing machines to enhance the smelting process.

Experiments using steam engines to power boats were being conducted in both the United Kingdom and America, and the first commercially successful river paddle steamer, Robert Fulton's Clermont, debuted in America in 1807. Then, in 1823, an Englishman named George Stephenson founded a locomotive company in Newcastle and, two years later, showed the viability of steam-powered railroad transportation.

Road construction techniques advanced throughout this time period. The most notable road builder of this era was Scotland's John Macadam

(1756–1836), who pioneered a method of road construction that included compacting layers of shattered stone. In addition to constructing around 180 miles of turnpike, Macadam also published many books on road construction.

Thomas Telford, a Scottish contemporary of Macadam, advocated creating roads using massive flat stones put on edge and jammed together to provide a stable basis that was paved with broken stone and gravel. During the early nineteenth century, Telford oversaw the building of over 920 miles of road and 1200 bridges.

One of the most significant engineering achievements of the nineteenth century was the discovery of electricity as a source of power. That accomplishment is primarily due to the work of many scientists and engineers in the latter part of the nineteenth century.

It was, however, founded on the discoveries of early-nineteenth-century physicists who established the fundamental nature of electricity: scientists like George Simon Ohm of Germany, Alessandra Volta of Italy, and Charles Coloumb and André Ampère of France. Machinery for new industries the eighteenth century also saw a rise in the status of engineering as a profession.

The first civil engineer, John Smeaton of the United Kingdom, was highly esteemed in scientific circles. In 1771, he assisted in the formation of an engineering organization with ambitions and traditions comparable to those of the Royal Society, of which he was a member.

The Institution of Civil Engineers was founded in 1818 by a group of young English engineers who chose Thomas Telford as its first president. George Stephenson was the first president of the Institution of Mechanical Engineers, which was founded in 1847.

1.1.7. Engineering in the Twentieth Century

Several key technical discoveries occurred during the first decade of the twentieth century that was destined to have a substantial influence on our civilization. At the turn of the century, inventors, and engineers were working feverishly to accomplish heavier-than-air flight. Success came in 1903 when Wilbur and Orville Wright flew their airplane for 12 seconds over a distance of 120 feet. Since that first flight, air transportation has evolved to dominate long-distance public carrier transit, accounting for 91% of intercity public passenger miles in the United States in 1998. Today, commercial airliners typically travel at speeds of 550 miles per hour, while supersonic aircraft may travel at speeds of up to 1450 miles per hour.

In the United States, about 3000 airports have been developed to support air travel, including over 2200 general aviation airports that handle smaller private planes. Atlanta Hartsfield International Airport handled more than 78 million incoming and leaving travelers in 1999. By 1900, a variety of "horseless carriages" had been developed, and by 1904, large numbers of motor vehicles were being produced.

Henry Ford made significant contributions to the development and appeal of vehicles by bringing contemporary mass manufacturing and low vehicle pricing. At the conclusion During the twentieth century, roughly 9 out of every 10 homes in the country had access to a motor car. More than half of all households in the United States possessed two or more motor cars.

To facilitate motor vehicle traffic, a 3.8-million-mile highway system has been built. The 45,500-mile Interstate Highway System, which cost more than $100 billion to build, is the most notable feature of this road network. This system, which was established in 1956, accounts for approximately 23% of all automotive distance traveled (Figure 1.3).

Figure 1.3. Former blast furnace in port of Sagunt, Valencia, Spain.

Source: Image by Wikipedia.

The Panama Canal, which opened in 1914, was the first massive building project completed in modern times. The canal is approximately 50 km long. It contains three lock sets, each with a length of 1000 feet, a width of 110 feet, and a depth of around 70 feet. A ship going from New York to San Francisco had to travel more than 13,000 kilometers around the tip of South America before it opened.

The canal cut that voyage down to around 5200 kilometers. Water resources were the subject of several twentieth-century technical feats. The Hoover Dam, which was built in 1936, is one example of this advancement. This pioneering concrete dam with a height of 726 feet was the world's tallest at the time of its completion.

The Tennessee Valley Authority's (TVA) impressive flood control, navigation, and power projects are another example of success in water resource management (TVA). The TVA, founded in 1933, delivered flood control, inexpensive power, and industrial prosperity to the Tennessee Valley. Shortly after World War II (WWII), design, and feasibility studies on nuclear energy generating were conducted in 1967, the first nuclear power station went into service.

Nuclear power has become economically competitive with fossil fuel power, and by 1998, 104 nuclear reactors in the United States were providing 674 billion kilowatt-hours of electricity, accounting for 21% of the nation's energy production. Heat is typically created in a commercial nuclear power-producing plant by the fission of a nuclear element such as uranium 235.

A steam generator removes the heat, and the steam is utilized to power a turbine and an alternator, which creates electricity. Nuclear power plant builders have encountered two significant obstacles. (1) establishing suitable precautions against radioactive emissions (e.g., adequate shielding, closed-cycle cooling systems); and (2) designing a protective containment structure to reduce the impacts of an explosion.

The twentieth century was marked by unprecedented technological advancement and upheaval. The increasing speed of discovery is likely most visible in the realm of electronics. In this century, rudimentary signal transmission has given way to contemporary communications networks with large switching systems based on electronic components.

Since the discovery of the transistor in 1947, semiconductor devices have largely supplanted vacuum tubes as electronic signal amplifiers. The transistor and semiconductor diode have significantly reduced the size of electronic equipment. The introduction of low-cost integrated circuits (ICs) mass-produced on small silicon chips has resulted in dramatic improvements in electrical design.

Along with downsizing, such devices have offered reliable and speedy signal transmission via circuits, resulting in the creation of faster switching circuits and digital computers. We have only been able to quickly outline a handful of the outstanding achievements of engineers over the twentieth century due to space constraints.

It is anticipated that the examples provided will help the reader understand some of the excitement and challenges connected with a career in engineering. Engineers will face a slew of complicated challenges with far-reaching consequences in the future, including:

- The discovery, development, and application of alternative energy sources to supplement the world's decreasing supply of coal and petroleum.
- Research and development of methods to preserve and repair the nation's huge failing public works infrastructure.
- Further advancement of microcomputer technology and its applications.
- Technological advancements to boost agricultural output in order to meet the world's expanding population and hunger.
- Creating structures that are more resistant to earthquakes, hurricanes, and other natural disasters.
- Improved methods for managing hazardous waste disposal, including radioactive waste generated during nuclear power generation.
- Interplanetary space exploration and the finding of space research applications for military and peaceful purposes.

1.2. DEFINITION OF ENGINEERING

Engineering is defined by the Accreditation Board for Engineering and Technology (ABET) as "the profession in which knowledge of the mathematical and natural sciences gained through study, experience, and practice is applied with judgment to develop ways to economically utilize the materials and forces of nature for the benefit of mankind."

Certain key aspects that explain the core of engineering are embodied in that definition. Engineering is a career. It aims to high standards of behavior and acknowledges duties to customers, peers, and society as a whole, just like law, medicine, architecture, education, and the ministry.

It is founded on a specific body of knowledge, and its members achieve professional status through well-defined educational and training pathways. Engineering is seen as both an art and a science. It is said to embody a set of concepts, procedures, and talents that cannot be gained only by study. It must be learned by experience and professional practice, at least in part.

Professional judgment must be used to moderate the engineer's expertise. Engineering challenges must fulfill competing objectives, and the recommended optimum solution does not necessarily emerge from a straightforward application of scientific ideas or formulas.

The engineer must balance competing limitations and make decisions based on knowledge and experience in order to find the best or optimum solution. Engineers use natural materials and forces to find solutions to issues. Engineers may use an almost infinite number of materials, both natural and man-made may use to create their ideas.

They choose materials based on their availability, pricing, and physical qualities (weight, strength, durability, elasticity, and so forth). The engineer has access to a considerably narrower range of energy sources, including petroleum, coal, gas, nuclear fission, hydroelectric power, sunshine, and wind.

The availability, cost, safety, and technological sophistication of these sources vary greatly. Engineers know that the earth's supply of materials and energy is finite, and therefore must be concerned not just with resource consumption but also with resource conservation.

This includes recycling and reusing existing resources, rehabilitating rather than replacing outdated facilities, and creatively substituting a plentiful item for one in short supply. It also entails looking for solutions that are energy efficient and looking for new energy sources to replace ones that are exhausted.

1.3. ENGINEERING FIELDS OF SPECIALIZATION

Engineering is a broad field. It is made up of numerous primary branches or areas of specialty, as well as dozens of lesser branches. Engineers established these fields in response to a growing body of technical knowledge. Some of the more notable branches of engineering are described in the following paragraphs.

These are addressed in general order of diminishing discipline size. It should be noted that there is significant overlap among the various specializations. During the course of a career, it is not unusual for an engineer to practice more than one specialty within a main discipline.

1.3.1. Electrical Engineering

Electrical engineering, the most extensive of all engineering disciplines, is concerned with electrical devices, currents, and systems. Electrical engineers deal with a wide range of equipment, from large power generators to microscopic computer chips.

Their work benefits nearly every area of society, including electrical appliances for households, electronic displays for businesses, lasers for industry, and satellite systems for government and enterprises. Electrical engineers are in charge of electricity generation, transmission, and distribution (Figure 1.4).

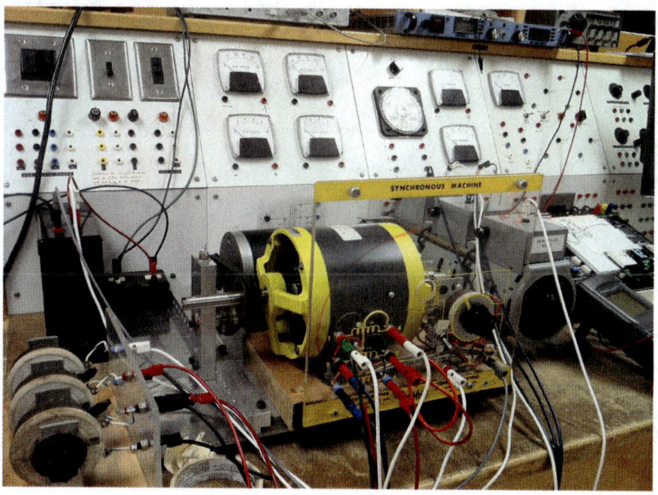

Figure 1.4. Electrical engineering test equipment.

Source: Image by Flickr.

They find hydroelectric, steam, diesel engine, and nuclear power facilities and specify their engines, generators, and ancillary equipment. These engineers have contributed to the establishment of electric-generating stations in the United States that have a capacity of around 700 million kilowatts and generate more than 3 trillion kilowatt-hours of electricity each year.

1.3.2. Computer Engineering

Computer engineering is the branch of engineering that is in charge of designing and implementing digital systems, as well as integrating computer technology into an expanding variety of systems and applications. It is a relatively young and fast emerging engineering subject that presents exceptional difficulties and potential (Figure 1.5).

Figure 1.5. Female engineer at computer.

Source: Image by Flickr.

Technological developments in computer engineering have been amazing since the introduction of the transistor in 1947. More than 100 million transistors may now be packed onto a single IC chip. Simultaneously, transistor switching speeds have grown more than 10,000 times, and it is now feasible to create devices that do more than one billion operations per second.

Because of the rapid advancement of computer technology, computer engineers have been pushed: (1) to innovate hardware and software design, as well as the tools used to construct these IC chips; (2) to envision, create, and test systems containing these chips.

Rapid improvements in computer technology have resulted in ever smaller, less expensive, high-performance computers, leading in a plethora of applications utilizing embedded computers as components. These include everything from very complicated communication systems to biological imaging gadgets, sophisticated consumer goods, and domestic appliances.

1.3.3. Mechanical Engineering

Mechanical engineering, one of the oldest and most diverse branches of engineering endeavor, is concerned with machines, power, and manufacturing or production processes. Mechanical engineers create machine tools—the machines that build machines—as well as machinery and equipment for all industries.

Turbines, printing presses, earth-moving machinery, food processors, air conditioning and refrigeration systems, prosthetic hearts and limbs, and engines for airplanes, diesel locomotives, automobiles, and trucks, and public transit vehicles are just a few examples (Figure 1.6).

Figure 1.6. A mechanical engineering student.

Source: Image by Flickr.

Their devices move and lift weights, carry people and things, and generate and convert energy. Mechanical engineers in the power specialty are involved in the design, manufacture, and operation of hydraulic turbines for driving electric generators, as well as boilers, engines, turbines, and pumps for the creation of steam power.

They develop and run power plants and are concerned with the efficient burning of fuels, the conversion of thermal energy into mechanical power, and the application of that power to practical work.

1.3.4. Civil Engineering

Many of our everyday activities are influenced by civil engineering: the structures we live and work in, the transportation facilities we use, the water we drink, and the drainage and sewage systems that are essential to our health and well-being.

The greatest subfield of civil engineering, structural engineering, is concerned with the design of major structures such as buildings, bridges, tanks, towers, dams, and other massive structures. These engineers develop and choose appropriate structural components and systems (e.g., beams, columns, and slabs) to offer necessary strength, stability, and longevity.

A significant number of civil engineers work in the construction business, erecting the structures that other engineers and architects design. Construction engineers are responsible for utilizing and managing construction resources (vehicles, equipment, machinery, materials, and experienced personnel) in order to build the structure or facility envisioned by the designer in a timely and efficient manner.

Transit engineers are responsible for the design and construction of roads, airports, harbors, and ports, and public transportation networks. They plan and develop transportation terminals, as well as invent and run controlling traffic systems Geotechnical engineers study the structural behavior of soil and rock (Figure 1.7).

Figure 1.7. Civil engineering students at Assam Don Bosco University in the state of Assam, India.

Source: Image by Wikimedia commons.

They investigate earth support systems and create foundations, earth barriers, and highway and airport pavements. Water flow through ditches, conduits, canals, dams, and estuaries is a challenge for hydraulic and water resource engineers. They build dams, irrigation systems, municipal water works, and drainage and erosion control systems using their unique knowledge of fluid mechanics.

Environmental engineers are concerned with waste management, air and water pollution, pesticide control, and radiation dangers. They develop and manage water treatment and sewage treatment plants, as well as to measure and monitor pollution in the air, on land, and in lakes and streams. Geodetic engineers are those who measure and map the earth's surface.

They precisely determine property and building lines and survey engineering project sites, elevations, and alignment. Civil engineers work for building firms, manufacturers, power providers, and consulting engineering organizations. There are several options for civil engineering jobs in local, county, and state engineering departments, as well as in federal government organizations.

1.3.5. Chemical Engineering

Chemical engineering is the use of chemistry, physics, and engineering in the design and operation of facilities for the creation of materials that undergo chemical changes throughout production (11). Paints, lubricants, fertilizers, medicines, cosmetics, petroleum products, foods, metals, polymers, ceramics, and glass are examples of such materials.

Chemical engineers are responsible for developing systems for generating vast amounts of materials that chemists generate in tiny quantities in the laboratory in these and other sectors. Chemical engineers pick appropriate procedures and sequence them to generate the desired output. Chemical engineers are in high demand in almost every industry.

In the future, there will be a strong and rising demand for chemical engineers as firms continue to create new goods in response to people's desire for improved health and a greater quality of life. Chemical engineers can also be anticipated to play a significant part in solving some of humanity's most difficult challenges, such as environmental degradation, depletion of energy supplies, and global overpopulation and famine.

1.3.6. Industrial Engineering

Industrial engineers work on the design, improvement, and installation of integrated systems of people, materials, and energy in the production of commodities or services. They design procedures and systems to increase quality and productivity. They are especially engaged in issues with the efficient use of money, materials, time, human labor, and energy.

They are more interested in the "big picture" of industrial management and production than in the details of process development. Industrial engineers conduct time and motion studies on workers, establish job performance standards, and suggest new and improved work practices to boost productivity. To prevent waste and client complaints, they use quality control procedures.

They employ statistical approaches to define realistic quality tolerances and design systems for doing routine product quality inspections. In all of their endeavors, Industrial engineers must closely monitor production expenses and look for methods to save costs without sacrificing product quality.

1.3.7. Aerospace Engineering

All elements of vehicle flying at all speeds and altitudes are addressed by aerospace engineering. It comprises hovercraft intended to function a few feet above land and water, helicopters that hover and maneuver in all directions, a range of conventional airplanes, and complicated spacecraft for circling the Earth and exploring the solar system.

Aeronautical engineering is concerned with atmospheric flight, whereas astronautical engineering is concerned with space travel. Aerospace engineers often specialize in one of many disciplines, including aerodynamics, structural design, propulsion systems, and guidance and control. Aerodynamics is concerned with the efficient design of the outside surfaces of aeronautical vehicles.

Aerodynamics engineers supervise wind tunnel tests, measure, and forecast lift and drag forces, and develop and test theories of flight performance, stability, and control. Structural designers in aeronautical engineering strive to design and manufacture aircraft systems that are cost-effective to operate.

This typically means optimizing the vehicle's strength-to-weight ratio. They also investigate how aircraft structures respond to mechanical vibrations and other dynamic forces, and they develop structures that can withstand these forces. Aerospace engineers design and improve aero plane and spaceship propulsion systems.

The thrust for all of the many propulsion systems for aerospace vehicles is created by accelerating a fluid backward. In the case of a turbojet, burning fuel creates hot gas, which is expanded further by a jet nozzle and provides thrust. Other systems, such as helicopters and tiny, low-speed aircraft, use a propeller powered by an engine that generates power by compressing, burning, and expanding its fuel (Figure 1.8).

Figure 1.8. Female aerospace engineer in hangar with the Tempest aircraft.

Source: Image by Flickr.

1.4. INTRODUCTION TO THE ENGINEERING PROFESSION

Engineers solve problems. Successful engineers are strong communicators and team players. They understand fundamental physical laws and math. Engineers use physical and chemical principles, as well as mathematics and mathematics, to design, develop, test, and manage the production of millions of goods and services.

When developing goods, they examine essential variables such as sustainability, efficiency, cost, dependability, and safety. Engineers are committed to lifelong learning and community service. Engineers create things and deliver services that improve our lives.

Look around you more closely tomorrow morning when you get up to notice how engineers contribute to the comfort and improvement of our daily life. Your bedroom was kept at the proper temperature during the night owing to mechanical experts who developed your home's heating, air-conditioning, and ventilation systems.

When you turn on the lights in the morning, know that thousands of mechanical and electrical engineers and technicians at power plants and power stations across the country are ensuring that the flow of electricity remains uninterrupted so that you have enough power to turn on the lights or turn on your TV to view the day's morning news and weather report.

Electrical and electronic engineers created the television (TV) you're watching to obtain your morning news. Of course, engineers from other disciplines are engaged in the final product's development, such as manufacturing and industrial engineers. When you're getting ready for your morning shower, the clean water you're going to use is being delivered to your home by civil and mechanical experts.

Even if you live on a farm in the country, the pump that brings water from the well to your house was developed by mechanical and construction engineers. Natural gas, which is transported to your home owing to the efforts of chemical, mechanical, civil, and petroleum experts, might be used to heat the water. When you're getting ready to dry yourself with a towel after your morning shower, consider the sorts of engineers that labored behind the scenes to create the towels.

Yes, agricultural, industrial, manufacturing, chemical, petroleum, civil, and mechanical engineers worked together to create the cotton towel. Consider the machines that were used to select the cotton, transport it to a factory, clean it, and dye it to an attractive hue to your eyes. Other machines were then utilized to weave the cloth and deliver it to sewing machines created by mechanical engineers.

1.5. ENGINEERING TECHNOLOGY

We introduced you to the engineering profession and its different areas of expertise in the preceding chapter. Let us now talk about engineering technologies. Engineering technology may be a good fit for those of you who are more hands-on and less interested in theory and mathematics.

Engineering technology programs often need fundamental mathematics up to the level of integral and differential calculus and focus on the application of technologies and processes. Engineering technologists, to a lesser extent than engineers, apply the same ideas of science, engineering, and mathematics to aid engineers in overcoming industrial challenges.

Construction, product creation, inspection, maintenance, sales, and research are just a few examples. They may also help engineers or scientists set up experiments, run tests, gather data, and calculate findings. In general, an engineering technologist's job is more application-oriented and needs less knowledge of mathematics, engineering theories, and scientific ideas required in sophisticated designs.

Engineering technology programs often include the same disciplines as engineering programs. You may, for example, earn a degree in Civil Engineering Technology. Mechanical Engineering Technology, Electronics Engineering Technology, or Industrial Engineering Technology are all examples of engineering technologies.

If you opt to pursue an engineering technology degree, keep in mind that graduate courses in engineering technology are limited, and registration as a professional engineer (PE) in some jurisdictions may be more challenging. The Accreditation Board for Engineering and Technology has also accredited the engineering technology programs (ABET).

ABET requires that Baccalaureate engineering technology programs have a minimum of 124 semester hours or 186 quarter hours of credit. Associate degree (two-year) programs must have at least 64 semester hours or 96 quarter hours of credit. Furthermore, any engineering technology program must have the following five components: communication, mathematics, physical and natural science, social sciences and humanities, and technical content.

A specific engineering technology program's technical content focuses on the applied side of science and engineering and is meant to acquire the skills, knowledge, methods, processes, and techniques connected with that particular technical discipline.

1.6. INTRODUCTION TO ENGINEERING DESIGN

Engineers, regardless of their expertise, follow a set of procedures when creating the goods and services we use every day.

These are the steps:

- Recognizing the need for a product or service;
- Fully defining and comprehending the problem (the need);
- Conducting preliminary research and preparation;
- Conceptualizing ideas for potential solutions;
- Synthesizing the results;
- Evaluating good ideas in greater detail;
- Optimizing the result to arrive at the best possible solution; and
- Presenting the solution.

1.7. ENGINEERING DESIGN PROCESS

Engineers develop millions of things and services that we use every day by applying physical laws, chemical laws and principles, and mathematics. Automobiles, computers, aircrafts, clothes, toys, household appliances, surgical equipment, heating, and cooling equipment, health care devices, tools, and machinery used to manufacture various items, and so on are examples of these products.

Engineers create goods with crucial criteria like cost, efficiency, dependability, sustainability, and safety in mind, and they test them to ensure that they can endure varied loads and situations. Engineers are always looking for new ways to improve old items.

Engineers also design and manage the construction of structures such as buildings, dams, roads, and public transportation networks. They also design and oversee the building of power plants that deliver energy to factories, residences, and businesses.

Engineers play an important role in the design and maintenance of a country's infrastructure, which includes communication networks, utilities, and transportation. They are always developing new innovative materials in order to make items lighter and stronger for various purposes.

Engineers are also in charge of devising appropriate methods and designing the essential equipment for extracting petroleum, natural gas, and raw minerals from the soil. Let us now take a closer look at what forms the design process. These are the fundamental processes that engineers, regardless of their expertise, take to solve issues (Figure 1.9).

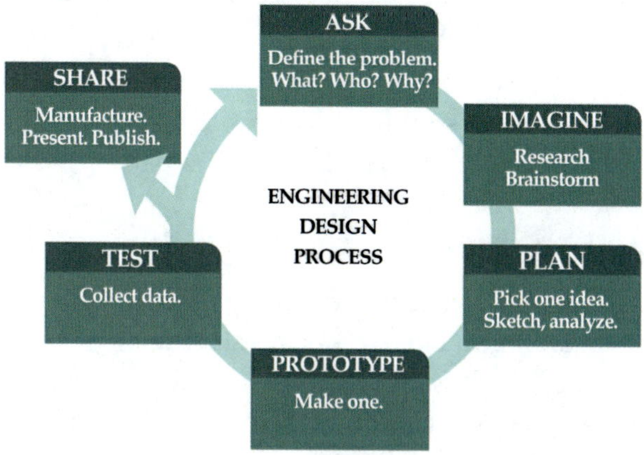

Figure 1.9. Engineering design process in six steps.

Source: Image by Wikimedia commons.

Keep in mind that the stages we will go through shortly are not independent of one another and do not necessarily follow one another in the sequence shown above. In reality, when clients decide to modify design specifications, engineers frequently need to return to stages 1 and 2.

Engineers are frequently obliged to provide oral and written status updates on a regular basis. As a result, while we included presentation of the design process as step 8, it might very well be an intrinsic aspect of many other design processes. Let us now examine each phase in further detail, beginning with the necessity for a product or service.

1.7.1. Recognizing the Need For a Product or a Service

All you have to do is chance to look around to see how many goods and services are built by engineers you use every day. Most of the time, we take these goods and services for granted until there is a disruption in the services they give. Some of these old items are regularly updated to take advantage of new technologies.

Automobiles and household appliances, for example, are always being updated to accommodate new technology. In addition to the items and services that are already in use, new products are being produced on a daily basis to make our lives more pleasant, joyful, and less tedious.

There's also the old adage that "every time someone complains about a circumstance, a chore, or a product, there's a potential for a product or a service right there." As you can see, there is a need for products and services; all that remains is to identify them.

The need may be discovered by you, your future employer, or a third-party customer who needs a solution to a problem or a new product to make what it does easier and more efficient.

1.7.2. Problem Definition and Understanding

As a design engineer, one of the first things you must do is completely comprehend the challenge. This is the most crucial phase in any design procedure. If you don't understand the problem or what the customer wants, you won't be able to provide a solution that is relevant to the client's needs. The greatest method to completely comprehend a situation is to ask several questions.

You may ask the customer, "How much money are you willing to spend on this project?" Is there a limit to the size or type of materials that may be used? When do you require the goods or services? How many of these items do you require? Inquiries frequently lead to other questions that help to identify the situation.

Furthermore, keep in mind that engineers typically operate in a team context, consulting with one another to solve complicated challenges. They break the work into smaller, manageable difficulties; as a result, effective engineers must be strong team players.

Because of the global economy, good interpersonal and communication skills are becoming increasingly vital. You must ensure that you understand your share of the problem and how it relates to the other difficulties. For example, separate businesses in different states or countries may manufacture distinct elements of a product.

Cooperation and coordination are required to guarantee that all components fit and perform effectively together, which necessitates good collaboration and strong communication skills. Before proceeding to the next phase, ensure that you understand the problem and that it is adequately described.

1.7.3. Research and Preparation

Once you have a thorough understanding of the situation, the following stage is to gather helpful information. In general, a smart place to start is by Googling to see whether a product that nearly fulfills the needs of your customer currently exists.

Perhaps your organization has already produced a product, or components of a product, that you might alter to fulfill the requirement. You certainly don't want to "reinvent the wheel!" As previously said, depending on the scale of the project, certain projects require collaboration with other organizations, therefore you must learn what is accessible through these other companies as well.

Try to get as much data as possible, this is where you will spend the most of your time, not just with the customer, but also with other engineers and technicians. Search engines on the internet are becoming increasingly crucial instruments for gathering such information. After gathering all relevant information, you must examine it and organized it appropriately.

1.7.4. Conceptualization

During this design phase, you should produce some ideas or concepts that might provide plausible answers to your problem. In other words, without conducting an extensive study, you must generate some potential solutions to the problem. You must be imaginative and come up with various alternate ideas.

You do not need to rule out any good workable notion at this level of design. If the problem involves a complicated system, you must identify the system's components. You should not have to go into the specifics of each prospective option just yet. However, you must conduct sufficient analysis to determine whether the thoughts you propose have validity.

Simply put, you must answer the following question: Would the principles work if they were developed further? You must learn to budget your time throughout the design process. Engineers with good time management abilities may operate successfully and efficiently.

You must learn how to make a milestone chart that details your timetable for finishing the project. You must display the time periods as well as the tasks that must be completed throughout these times.

1.7.5. Synthesis

Recollect from Chapter 1 that competent engineers understand the underlying principles of engineering and can use them to solve a wide range of issues. Engineers must be analytical, detail-oriented, and imaginative. During this stage of design, you start thinking about specifics.

You must conduct calculations, run computer models, narrow down the sort of materials to be utilized, size the system components, and answer questions about how the product will be manufactured. You will study relevant codes and standards to ensure that your design complies with these rules and standards.

1.7.6. Evaluation

Examine the issue in further depth. It is possible that you may need to identify crucial design characteristics and assess their impact on your final design. At this point, you must ensure that all computations are carried out accurately. If your analysis has any uncertainties, you must do an experimental inquiry.

Working models must be produced and tested wherever feasible. The optimal solution among the alternatives must be determined at this stage of the design approach. The details of how the product will be manufactured must be thoroughly thought out.

1.7.7. Optimization

Optimization can refer to either minimization or maximization. A functional design and an optimal design are the two primary forms of design. A functional design is one that fits all of the predetermined design standards while also allowing for improvement in some areas. Consider an example to better grasp the notion of functional design.

Assume we are designing a 3-meter-tall (10-foot) ladder to hold a person weighing 1335 newtons (300 pounds) with a certain safety factor. We will devise a design that includes a steel ladder that is 3 meters (10 feet) tall and can safely hold a load of 1335 N (300 lb) at each step. A specific amount of money would be required to purchase the ladder.

This design would meet all of the requirements, including those for strength and size, and would thus be considered functional. Before we can think about enhancing our design, we need to figure out what criterion we should apply to optimize it. Design optimization is always driven by a certain criterion, such as price, strength, size, weight, dependability, noise, or performance. If we utilize weight as an optimization criterion, the challenge becomes decreasing the weight of the ladder while maintaining its strength. For example, we could make the ladder out of metal. We would also complete stress analysis on the new ladder to evaluate if we could eliminate material from certain areas of the ladder without impacting loading and safety regulations.

Another thing to remember is that improving individual components of an engineering system does not always result in an optimal whole. Consider a thermal-fluid system, such as a refrigerator. Individual component optimization for some criterion, such as the compressor, evaporator, or condenser, does not result in an optimized overall system (refrigerator).

1.7.8. Presentation

You must now explain your final answer to the client, who might be your supervisor, another group inside your firm, or an outside customer. You may be required to produce both an oral presentation and a written report. Engineers are expected to prepare reports, depending on the nature of the project, these reports may be extensive, thorough technical reports with graphs, charts, and engineering drawings, or they could be concise memorandums or executive summaries.

Although we have included the presentation as Step 8 of the design process, engineers are frequently expected to deliver oral and written status updates on a regular basis to a variety of organizations, as a result, presentation may become an essential component of many other design phases. We have devoted a whole chapter on engineering communication due to its relevance.

Engineers also have strong "people skills," which allow them to engage and communicate successfully with a variety of individuals inside their business. For example, they can interact effectively with both sales and marketing professionals and their own engineering colleagues.

1.8. CONCLUSION

In the conclusion of this chapter, it discussed about the history of engineering. This chapter also provides highlights on the definition of engineering. This chapter also explains various engineering fields of specialization. In this chapter, it also discussed about the engineering technology. Towards the end of the chapter, it addresses the several engineering design process, and its significance.

REFERENCES

1. Saeed Moaveni, (2011). *Engineering Fundamentals*. [e-Book] https://www.hzu.edu.in. Available at: https://www.hzu.edu.in/engineering/Engineering%20Fundamentals.pdf (accessed on 1 July 2022).
2. Agarwal, A., Kumar, D., Sharma, N., & Sonawane, U., (2021). Introduction to engine modeling and simulation. *Energy, Environment, and Sustainability* (pp. 3–6). [online] Available at: https://link.springer.com/chapter/10.1007/978-981-16-8618-4_1 (accessed on 1 July 2022).
3. Fadli, Z., (n.d). *[Paul H Wright] INTRODUCTION to Engineering(BookZZ.org)*. [online] Academia.edu. Available at: https://www.academia.edu/30141729/_Paul_H_Wright_Introduction_to_engineering_BookZZ_org_ (accessed on 1 July 2022).
4. Kiran, D., (2022). Introduction to engineering economics. *Principles of Economics and Management for Manufacturing Engineering* (pp. 3–10). [online] Available at: https://www.sciencedirect.com/science/article/pii/B9780323998628000273?via%3Dihub (accessed on 1 July 2022).
5. Panchangam, S., (2015). *Introduction to Engineering*. [online] https://www.researchgate.net. Available at: https://www.researchgate.net/publication/280302115_An_Introduction_to_Engineering/link/55b06ac908ae32092e0709f5/download (accessed on 1 July 2022).
6. Saterbak, A., & Wettergreen, M., (2021). Introduction to engineering design. *Synthesis Lectures on Engineering, Science, and Technology*, 3(3), i–223. [online] Available at: https://www.morganclaypool.com/doi/10.2200/S01095ED1V01Y202104EST016 (accessed on 1 July 2022).

CHAPTER 2

ENGINEERING AS A PROFESSION: AN OVERVIEW

CONTENTS

People, regardless of where they reside, require the essential necessities: food, clothes, shelter, and water for drinking or cleansing. Furthermore, we require numerous forms of transportation to go to other locations since we may live and work in different cities or desire to visit friends and family who reside elsewhere.

We also want to feel secure, to be able to rest and be entertained. We must also be loved and valued by our friends and family. More is required of engineers as a result of global socioeconomic demographic trends, environmental concerns, and the earth's finite resources.

Future engineers are expected to produce goods and services that improve living standards and health care while simultaneously addressing severe environmental and sustainability challenges. At the turn of the twentieth century, there were around six billion of us on the planet.

2.1. INTRODUCTION

As a point of reference, the world population was one billion people 110 years ago, at the turn of the nineteenth century. Consider this. It has taken us from the dawn of time to achieve a population of one billion. It just took 110 years to more than fivefold the population. Some of us enjoy a high level of life, while others in underdeveloped nations do not.

You would most likely agree that our world would be a better place if everyone of us had enough to eat, a nice and safe home to live, important job to perform, and time to rest. According to the most recent US Census Bureau estimates and predictions, the world population will reach 9.3 billion people by 2050.

Not only will the global population continue to grow, but so will the age structure of the global population. In the next 25 years, the world's old population (those over the age of 65) will be more than double.

2.2. WHAT IS ENGINEERING AND WHAT DO ENGINEERS DO?

Engineers employ physical and chemical laws and principles, as well as mathematics, to create millions of goods and services that we use every day. Automobiles, computers, airplanes, clothing, toys, household appliances, surgical equipment, heating, and cooling equipment, health care devices, tools, and machinery used to manufacture various items, and so on are examples of these products.

When creating these goods, engineers evaluate critical variables such as cost, efficiency, dependability, and safety. Engineers conduct tests to ensure that the goods they develop can endure a variety of loads and circumstances. They are always looking for new methods to improve old items.

They also design and manage the construction of buildings, dams, roads, and public transit systems, as well as power plants that give power to factories, residences, and offices. Engineers play an important part in the design and upkeep of a country's infrastructure, which includes communication networks, public utilities, and transportation (Figure 2.1).

Figure 2.1. National engineers' week: NAVFAC pacific fire protection engineer Jordan Lau inspects the fire sprinkler riser assembly.

Source: Image by Flickr.

2.3. ENGINEERING AS A PROFESSION AND COMMON TRAITS OF GOOD ENGINEERS

In this part, we will first explore engineering in general, and then we will narrow our emphasis to specific elements of engineering. We'll also look at the features and attributes that many engineers share. Following that, we will go through some specific engineering specialties.

As we mentioned previously in this chapter, some of you may not have determined what you want to study during your college years and, as a result, you may have numerous questions, including: What is engineering and what do engineers do? What are some of the engineering specializations? Is it true that I want to study engineering? How will I know whether I've chosen the ideal field for me?

Will there be a big demand for my area of concentration after I graduate? and what comes after that? The parts that follow are designed to assist you in making an informed decision; nevertheless, don't stress about obtaining answers to all of these issues right now.

You have some time to think about them because the majority of the curriculum during the first year of engineering is the same for all engineering students, regardless of field. So, you have at least a year to examine your options. This is true in the majority of educational institutions.

Nonetheless, you should consult with your counselor early on to establish how soon you must pick a field of expertise. Also, don't be anxious about your chosen career altering in such a manner that your degree becomes obsolete. To catch up with new technology, most organizations aid their engineers in obtaining further training and education.

A successful engineering education will prepare you to be a good problem solver for the rest of your life, regardless of the problem or scenario. You may ask why you need to understand some of the things you are studying throughout the next several years of school.

Your assignment may appear unimportant, inconsequential, or out of date at times. You may be confident that you are acquiring both topic material and thinking and analyzing methods that will prepare you to meet future issues that do not yet exist.

2.4. COMMON TRAITS OF GOOD ENGINEERS

Although engineers' tasks are fairly diverse, there are basic personality qualities and work habits that characterize the majority of today's successful engineers.

- Engineers are problem solvers.
- Good engineers understand the fundamental principles of engineering and can apply them to a wide range of issues.
- Good engineers are analytical, detail-oriented, and creative, and they want to be lifelong learners. For example, to keep current on

advances and new technology, they attend continuing education programs, seminars, and workshops.

- This is especially vital in today's environment since the quick developments in technology will demand you to stay up with new technologies as an engineer. Furthermore, if you do not constantly improve your engineering knowledge, you risk getting laid off or refused promotion.

- Good engineers, regardless of expertise, have fundamental knowledge that can be utilized in a variety of situations. As a result, well-trained engineers can work in domains other than their area of specialty. A good mechanical engineer, for example, with a broad base of knowledge, may work as an automobile engineer, an aerospace engineer, or a chemical engineer.
- Good engineers have written and spoken communication skills that enable them to collaborate effectively with their colleagues and communicate their knowledge to a diverse variety of clients.
- Good engineers have good time management abilities, which allow them to work successfully and efficiently.
- Engineers with high "people skills" can engage and communicate successfully with a variety of individuals in their business. They may, for example, interact effectively with sales and marketing specialists as well as their own colleagues.
- Engineers must produce written reports. These papers may be long technical reports with graphs, charts, and engineering drawings, or they could be concise memos or executive summaries.
- Engineers are skilled in modeling and analyzing diverse practical issues using computers in a variety of ways.
- By attending seminars, workshops, and meetings, good engineers actively participate in local and national discipline-specific organizations. Many of them even give talks at professional gatherings.
- Engineers often operate in a team atmosphere, consulting with one another to overcome difficult challenges They break the work into smaller, manageable difficulties; as a result, effective engineers must be strong team players.

For example, different businesses in different nations may manufacture different elements of an automobile. Cooperation and coordination are required to guarantee that all components fit and perform effectively together, which necessitates good communication skills.

These are some more engineering facts to consider.

- A bachelor's degree in engineering is necessary for practically all entry-level engineering positions.
- Starting pay for engineers are much greater than those for bachelor's degree holders in other industries, according to the US Bureau of Labor Statistics.
- Engineering has a very bright future. Between 2010 and 2018, new engineering graduates should have plenty of job prospects.
- The majority of engineering degrees are awarded in electrical, mechanical, and civil engineering, which are the forefathers of all other engineering fields.
- Engineers held 1.6 million employment in 2008.

2.5. WHAT ARE SOME AREAS OF ENGINEERING SPECIALIZATION?

Professional engineering societies recognize more than 20 primary fields or specialties. Furthermore, each discipline has a number of branches. Mechanical engineering programs, for example, are generally separated into two major areas: (1) thermal/fluid systems; and (2) structural/solid systems.

During your senior year in most mechanical engineering programs, you can select optional classes that allow you to follow your interests and enhance your knowledge base in these areas. For example, if you want to learn more about how buildings are heated or cooled in the winter, you will register in a heating Class of air conditioning systems.

Consider civil engineering for further information on the numerous branches within particular engineering fields. A civil engineering program's primary disciplines are typically environmental, geotechnical, water resources, transportation, and structural. Power generation and transmission, communications, control, electronics, and integrated circuits (ICs) are all examples of electrical engineering (EE) fields (Figure 2.2).

Figure 2.2. MECS workshop.

Source: Image by Wikimedia commons.

Although not all engineering disciplines are covered here, you are urged to visit the websites of relevant engineering societies to learn more about a certain engineering technologies and innovation.

2.6. PROFESSIONAL ENGINEER

Engineers whose job may jeopardize public safety must be registered in all 50 states and the District of Columbia. A degree from an ABET-accredited engineering program is required as the first step toward becoming a registered professional engineer (PE). During your senior year, you must also take the Fundamentals of Engineering Exam (FE).

The exam lasts roughly eight hours and is divided into two sections: morning and afternoon. You will answer multiple-choice questions in chemistry, physics, mathematics, mechanics, thermodynamics, electrical, and electronic circuits, and materials science throughout the morning session.

You will answer multiple-choice questions pertaining to your subject during the four-hour afternoon session, or you may choose to take a generic engineering test. After qualifying your FE test, you must obtain four years of relevant engineering job experience and pass a state-mandated eight-hour exam (the Principles and Practice of Engineering Exam).

Candidates select one of 16 engineering fields to take an exam in. Some engineers are licensed in multiple states. Professional registrations are often sought by civil, mechanical, chemical, and electrical engineers. You should anticipate to work under the supervision of a more experienced engineer as a recent engineering graduate.

Some firms may require you to attend workshops (short courses that can last a week) or a day-long seminar to gain further training in communication skills, time management, or a specific engineering approach based on your given job. As your expertise and experience grow, you will be allowed greater leeway in making technical choices.

After gaining extensive experience, you may choose to become a manager in control of a team of engineers and technicians. Some engineers begin their careers in sales or marketing connected to engineering products and services rather than in a specialized field of engineering.

As previously stated, professional bodies acknowledge more than 20 engineering fields. Most engineering degrees, however, are awarded in civil, electrical, and mechanical engineering. As a consequence, these fields are addressed first.

2.6.1. Civil Engineering

Civil engineering is one of the most ancient technical disciplines. Civil engineering, as the name indicates, is involved with the provision of public infrastructure and services. Buildings, roads, and highways, bridges, dams, tunnels, public transport systems, and airports are all designed and built by civil engineers. They also work on the design and monitoring of municipal water and sewage systems.

The important fields of civil engineering include structural, environmental, transportation, water resources, and geotechnical engineering. Civil engineers operate as consultants, construction supervisors, city engineers, public utility and transportation engineers, and city planners (Figure 2.3).

Figure 2.3. Female civil engineer.

Source: Image by Wikimedia commons.

According to the Bureau of Labor Statistics, the career prognosis for civil engineering graduates is favorable because as the population expands, more civil engineers will be required to design and supervise the construction of new buildings, roads, and water supply and sewage systems.

They are also required to supervise the upkeep and restoration of existing public structures such as highways, bridges, and airports.

2.6.2. Electrical and Electronic Engineering

The largest engineering discipline is electrical and electronic engineering. Electrical engineers design, develop, test, and supervise the production of electrical equipment such as lighting and wiring for buildings, cars, buses, trains, ships, and aircrafts; power generation and transmission equipment for utility companies; electric motors used in a variety of products; control devices; and radar equipment.

Power production, transmission, and distribution, and controls are the three primary fields of EE (Figure 2.4).

Figure 2.4. Image showing electronic board.

Source: Image by pixabay.

Electronic engineers are responsible for the design, development, testing, and supervision of electronic equipment, including computer hardware, computer network hardware; communication devices such as cellphones, TVs, and audio and video equipment; and measurement tools Computer and communication electronics are expanding disciplines of electronic engineering.

Electrical and electronic engineers have a bright future since corporations and governments want quicker computers and better communication networks. Of course, consumer electronics will have a big impact on employment growth for electrical and electronic engineers.

2.6.3. Mechanical Engineering

Mechanical engineering is one of the most diverse engineering fields, having changed through time as new technologies appeared. Mechanical engineers design, develop, test, and manufacture machines, robotics, tools, and power generation equipment such as steam and gas turbines, heating, cooling, and refrigerating equipment, and internal combustion engines.

Thermal/fluid systems and structural/solid systems are the two primary disciplines of mechanical engineering. Mechanical engineers have quite a promising work outlook too though as more efficient machinery and power generation equipment, as well as alternative energy-generating gadgets, are required.

Mechanical engineers can be found working for the federal government, consulting organizations, numerous manufacturing sectors, the automobile industry, and other modes of transportation.

Aerospace engineering, biomedical engineering, chemical engineering, environmental engineering, petroleum engineering, nuclear engineering, and materials engineering are some of the many popular engineering fields.

2.6.4. Aerospace Engineering

Commercial and military airplanes, helicopters, spaceships, and missiles are designed, developed, tested, and manufactured by aerospace engineers. They might be assigned to projects involving the research and development of guidance, navigation, and control systems.

The vast majority of aerospace engineers work for aircraft and missile manufacturers, as well as the Department of Defense and National Aeronautics and Space Administration (NASA). If you want to be an aerospace engineer, you should expect to reside in California, Washington, Texas, or Florida because these are the states with the most aerospace manufacturing enterprises.

According to the Bureau of Labor Statistics, the job outlook for aerospace engineers will be less favorable through 2010. One explanation for this slower job growth is a decrease in Defense Department spending. Commercial airplane makers, on the other hand, are likely to do well due to population increase and the need to fulfill the need for increasing passenger air traffic.

2.6.5. Biomedical Engineering

Biomedical engineering is a new discipline that combines biology, chemistry, medicine, and engineering to address medical and health-related issues. They build prosthetic limbs, organs, imaging systems, and medical gadgets using the laws and concepts of chemistry, biology, medicine, and engineering.

They also do research with medical professionals, chemists, and biologists to better understand biological processes and the human body. In addition to their biology and chemistry education, Mechanical or EE is a prerequisite for biomedical engineers (Figure 2.5).

Figure 2.5. A biomedical engineering laboratory.

Source: Image by Wikimedia commons.

Biomedical engineering has several specialties, including biomechanics, biomaterials, tissue engineering, medical imaging, and rehabilitation. Computer-assisted surgery and tissue engineering are two of the most rapidly increasing fields of biomedical engineering study.

Because of the emphasis on health concerns and the aging population, the job prospects for biomedical engineering graduates are quite strong, according to the Bureau of Labor Statistics.

2.6.6. Chemical Engineering

Chemical engineers, as the name indicates, apply the concepts of chemistry and fundamental engineering sciences to tackle a wide range of issues linked to the manufacturing of chemicals and their usage in a number of sectors, including pharmaceutical, electronic, and photographic industries.

Chemical, petroleum refining, film, paper, plastic, paint, and other related sectors employ the majority of chemical engineers. Chemical engineers are also employed in the sectors of metallurgy, food processing, biotechnology, and fermentation (Figure 2.6).

Figure 2.6. Chemical engineers talk in laboratory.

Source: Image by Flickr.

They often specialize in topics such as polymers, oxidation, fertilizers, and pollution control. According to the Bureau of Labor Statistics, the job prognosis for chemical engineers is likewise promising in order to meet the requirements of the increasing population.

2.6.7. Environmental Engineering

Another new field that has emerged as a result of our concern for the environment is environmental engineering. Environmental engineering, as the name suggests, is concerned with environmental issues. They use chemistry, biology, and engineering laws and concepts to handle difficulties like as water and air pollution management, hazardous waste, waste disposal, and recycling.

If these concerns are not handled effectively, they will have an impact on public health. Many environmental engineers work on the creation of municipal, national, and worldwide environmental laws and regulations. They investigate the impact of industrial and automotive emissions on acid rain and ozone depletion.

They also concentrate on issues related to the removal of existing hazardous material. Environmental engineers work as consultants or for government agencies at the municipal, state, and federal levels (Figure 2.7).

Figure 2.7. BYU environmental engineering lab.

Source: Image by Wikimedia commons.

The career future for environmental engineering graduates is extremely excellent, according to the Bureau of Labor Statistics, because environmental engineers will be needed in larger numbers to handle and regulate the environmental challenges outlined above.

It is vital to note that, more than other disciplines among engineers, the career future for environmental engineers is influenced by politics. Looser environmental policies, for example, might result in fewer jobs, whilst stronger policies may result in more jobs.

2.6.8. Manufacturing Engineering

Production engineers design, manage, and supervise the manufacturing process for various sorts of products. They are focused with producing items in an efficient and cost-effective manner. Manufacturing engineers work on all areas of production, such as scheduling and material handling, as well as the design, development, monitoring, and control of assembly lines.

For production goals, manufacturing engineers use robots and machine-vision technology. Manufacturing engineers construct prototypes of items before going to manufacture genuine products to illustrate concepts for new products and to save time and money.

This method is known as prototyping. Manufacturing engineers work in a variety of sectors, such as automotive, aerospace, and food processing and packaging. Manufacturing engineers have a promising career outlook.

2.6.9. Petroleum Engineering

Petroleum engineers are experts in the exploration and production of oil and natural gas. Petroleum engineers scour the world for subterranean oil or natural gas deposits in conjunction with geologists. Geologists are well-versed in the qualities of the rocks that make up the earth's crust.

Geologists collaborate with petroleum engineers to establish the optimum drilling methods after evaluating the attributes of the geological formations surrounding oil and gas deposits. Petroleum engineers also oversee and supervise drilling and oil extraction techniques.

Petroleum engineers create equipment and methods in partnership with other specialist engineers to obtain the most lucrative oil and gas recovery. As they experiment with different recovery approaches, they employ computer models to mimic reservoir performance.

If you choose to pursue a career in petroleum engineering, you will most likely work for one of the large oil firms or one of the hundreds of smaller, independent enterprises active in oil exploration, production, and servicing. Petroleum engineers are also employed by engineering consulting businesses, government organizations, oil field services, and equipment suppliers.

According to the US Department of Labor, petroleum engineers are employed in great numbers in Texas, Oklahoma, Louisiana, Colorado, and California, including offshore locations. Many American petroleum engineers also work in oil-producing countries like as Russia, the Middle East, South America, and Africa.

Petroleum engineers' career prospects are influenced by oil and gas prices. Despite this, if you opt to study petroleum engineering, career possibilities for petroleum engineers should be advantageous due to the historically low number of degrees given in petroleum engineering.

Furthermore, petroleum engineers work all across the world, and many businesses prefer US-trained petroleum engineers for positions in other nations.

2.6.10. Nuclear Engineering

Nuclear engineering programs are only available at a few engineering institutions around the country. Nuclear engineers design, build, monitor, and operate nuclear power equipment that uses nuclear energy to generate electricity. Nuclear engineers work on the design, development, and operation of nuclear power reactors that produce energy or power Navy ships and submarines.

They may also operate in sectors such as nuclear fuel manufacture and management, as well as the safe disposal of its waste products. Some nuclear engineers work on the creation of industrial and diagnostic medical devices. Nuclear engineers work for the United States Navy, nuclear power utilities, and the Department of Energy's Nuclear Regulatory Commission.

There are just a few nuclear power facilities under development due to the exorbitant cost and widespread public safety concerns. Nonetheless, the career prognosis for nuclear engineers is favorable because there are currently few graduates in this profession.

Nuclear engineers can also work in the Departments of Defense and Energy, as well as in nuclear medical technologies and nuclear waste management.

2.6.11. Mining Engineering

There are just a handful mining engineering colleges in the United States. Mining engineers work alongside geologists and metallurgical engineers to locate, remove, and prepare coal for use by utility companies. They also seek for metals and minerals to extract from the soil for use by various industrial businesses.

Mining engineers plan and manage the development of both aboveground and subterranean mines. Mining engineers may also be involved in the development of new mining equipment for mineral extraction and separation from other materials that are mixed in with the target minerals.

The majority of mining engineers work in the mining sector, but others work for government agencies or in the manufacturing business. Mining engineers do not have as excellent an employment outlook as other specialties. The mining sector is comparable to the oil industry in that job prospects are directly related to metal and mineral prices.

If the prices of these items are low, mining companies will be unwilling to invest in new mining equipment and mines. Mining engineers in the United States, like petroleum engineers, may find attractive prospects outside the country.

2.6.12. Materials Engineering

Only a few engineering schools have a formal curriculum in materials engineering, ceramic engineering, or metallurgical engineering. Materials engineers do research, design, and testing on novel materials for a variety of goods and technical applications. These novel materials may be metal alloys, ceramics, polymers, or composites.

Materials engineers research the composition, atomic structure, and thermo-physical characteristics of materials. They modify the atomic and molecular structure of materials to make them lighter, stronger, and more durable.

They develop materials with specific mechanical, electrical, magnetic, chemical, and heat-transfer properties for use in specific applications, such as graphite tennis racquets that are lighter and stronger than traditional wooden racquets; composite materials used in stealth military planes with specific electromagnetic properties; and ceramic tiles on the space shuttle that protect it during re-entry into the atmosphere (ceramics are nonmetallic materials that can withstand high temperatures).

Materials engineering can be subdivided further into metallurgical, ceramics, plastics, and other specializations. Materials engineers can be found working in aircraft production, research, and testing labs, and electrical, stone, and glass product makers. Because of the limited number of current graduates, work prospects for materials engineers are plentiful.

2.7. PREPARING FOR AN ENGINEERING CAREER

Transitioning from high school to college necessitates extra work. To have a satisfying education, you must recognize that you must begin studying and preparing on the first day of class, attend class consistently, seek help immediately, take good notes, choose one decent location of the study, and develop study groups.

To create a suitable weekly calendar, you should also consider the time management strategies described in this chapter. Your education is a costly investment. Make sensible investments.

2.7.1. Making the Transition from High School to College

You are now a member of an elite group of students since you are studying to be an engineer. According to the Chronicle of Higher Education, just around 5% of students who receive a B.S. degree from a university or college in the United States are engineers. You will be trained to perceive your environment differently than others.

You will learn how to ask questions to discover how things are built, how they operate, how to enhance them, how to design something from start, and how to take an idea from paper to reality and really construct something. Some of you may have been on your own for the first time. Making the move from high school to college may be a significant step for you.

Remember that what you do in the following four or five years will affect you for the rest of your life. Remember that your success and happiness are entirely dependent on you. Nobody can force you learn; you must take responsibility for it. Depending on whatever high school you attended, you may not have had to work very hard to earn decent grades.

The majority of your learning in high school occurred in the classroom. As a result, in order to prosper as an engineering student, you may need to acquire some new behavior and break some old ones. The remainder of this chapter provides advice and ideas to help you have a good college experience. Consider these ideas and attempt to modify them to your specific scenario.

2.7.2. Budgeting Your Time

Each of us has the same 24 hours in a day, and there is only so much that a person can achieve on an average daily basis. Many of us require about 8 hours of sleep every night. Furthermore, we all require time for work, friends, and family, learning, leisure, and entertainment, and just having fun.

Assume you were handed a million dollars as an adult and informed it was all the money you would have for the rest of your life to spend on clothing, food, entertainment, leisure activities, and so on. What are your plans for spending the money? Of course, you'd make reasonable attempts to spend and invest your money sensibly.

You would meticulously budget for various necessities, attempting to get the most bang for your buck. You would hunt for good deals and buy only what was necessary, attempting not to spend any money. Consider your education in the same way. Don't just pay your tuition and sit in class daydreaming.

Your education is an expensive investment that demands careful management on your behalf. A student at a private university addressed their instructor about leaving a class since they were not earning the appropriate grade. The teacher asked how much they'd paid for the training. They claimed to have spent roughly $2000 on the four-credit course.

The instructor happened to have a laptop computer on their desk and asked, "Would you throw away a laptop computer if you bought it from a computer store, took it home to install some software on it, and had some difficulty getting the device to work?" The students looked at their teacher as though they had posed a silly question.

They emphasized that abandoning a class for which they had already paid tuition is analogous to tossing away a computer the first time they had a problem with some software. Make an effort to learn from this example. In general, learning is a lot of effort for most of us at first, and it's not much pleasure.

However, within a short amount of time, studying may become a delight, something you focus on to boost your own self-esteem. Learning and comprehending new concepts may be exhilarating. Let us look at what you can do over the next several years to improve your learning and make the engineering education you are going to get a joyful and gratifying experience.

2.7.3. Daily Studying and Preparation

You begin learning and preparing for class the first day and! It is usually a good idea to read ahead of time the subject that your lecturer intends to discuss in class. This practice will help you learn and remember the course contents better. It is also critical to review the content that was addressed in class later that day following the lecture.

Reading the material ahead of a lecture allows you to become acquainted with the information that the lecturer will offer to you in class. Don't be concerned if you don't fully comprehend all you're reading at the moment.

During the lecture, you can ask questions on subject that you did not fully comprehend. Everything should come together when you go through the content after the lecture. Take some time to read before class and study the topic thereafter on the same day.

2.7.4. Your Graduation Plan

Most schools have three stages of admission. First and first, you must get admitted to the institution. Certain conditions must be met in order for this to occur. For example, you must be in the top x percent of your high school class, have a particular ACT or SAT score, and have completed a certain number of years of English, mathematics, science, and social studies.

After your freshman year, you may need to apply to the institution that houses the engineering school of your choice. You must fulfill extra standards to be admitted to, say, your university's engineering college. Finally, at the conclusion of the second year, after successfully completing algebra, chemistry, physics, and fundamental engineering studies, you must apply and be accepted.

Make an appointment with your adviser to discuss the prerequisites for admission to the college and the specific program, as admittance to an engineering school is very selective at many colleges. It is also advisable to meet with your adviser and plan your graduation. List all of the classes you'll need to take in order to graduate in four or five years.

You can always alter your mind later if your interests change. Make sure you understand the prerequisites for each subject and when a class is normally given; a program flowchart will come in handy. To raise your awareness of your societal duties as an engineer, you must also take a specified amount of social science and humanities classes.

Consider your present interests as well as your social science and humanities courses. Again, don't be concerned if your hobbies change; you can always change your strategy. Those of you who are presently enrolled in a community college and intend to transfer to a university should contact the institution to learn about their engineering course transfer regulations and requirements, and plan your graduation appropriately.

2.8. THE ENGINEERING PROFESSION AND THE ENGINEER OF THE 21ST CENTURY

We can't predict what the engineering profession will look like in a hundred years. The heated debates presently taking place among profession leaders and educators imply that innovation will be a prominent subject. The concept is that expertise is a commodity, and regular engineering services will be available from low-cost vendors anywhere in the globe.

As a result, engineering education must provide value beyond simply teaching skills. Of course, just because a talent is or will become a commodity does not mean that future engineers do not need to acquire them. On the contrary, they will need to be much more technically adept than individuals who currently make a career performing tightly specified jobs.

Engineers in the twenty-first century must be able to continually acquire information and decide on a course of action, including what tools are required for a certain task. How can we teach someone who is only a teenager to have these qualifications? Or, for that matter, do such broad remarks imply anything specific?

Our view is that they do, and that these aims, first, and foremost, translate into precise curricular needs, and secondly, the first aim, knowing anything, is rather simple. We can now "Google" any topic and almost certainly get a wealth of knowledge in a couple of seconds. And as search engines improve, the likelihood that the material is relevant will rise.

The transformational power of having quick access to knowledge cannot be overstated. We all "know more than we know" because, in addition to having knowledge, we also know where to get information about certain topics. Most of us know how to fix our computers, not because we know how, but because we know who to ask.

The emergence of the internet broadened this network of relationships to include practically every piece of information available. However, while accessing information is now simple, the communalization of knowledge will necessitate the PE's ability to appraise the quality of the information that he or she possesses.

Thus, the educational challenge will be to teach students how to deal with a plethora of information and how to determine the relevance and quality of the information at hand. Engineers have always learnt by taking on new difficulties. However, the expansion in the availability of tools to perform practically anything suggests that engineering instructors must reconsider how students are trained in the fundamentals of their fields.

Computer programs that can accomplish almost anything, from simple calculations to simulating complicated systems to designing a whole manufactured item, enable the modern engineer to achieve more than his or her forebears could ever conceive.

These tools, however, necessitate not only that the engineer understand how to use them, but also that he or she be able to judge whether tool is suited for a certain work and then critically analyze the output. "To err is human, but to truly screw up you need a computer," therefore common sense will be much more important when design and analysis are done entirely on the computer.

While teaching engineering students how the physical world works is still at the heart of engineering education, it is time to reconsider how we teach the principles of engineering science to students. Knowing the magnitude of phenomena and distributing knowledge across several scales are key skills.

Engineers' non-technical professional abilities must be adapted to the new manner of performing engineering, in addition to changes in the technical skills they must possess. In the United States, significant progress has already been made toward making communication in its broadest sense an intrinsic element of the engineering curriculum.

Most programs now demand graduates to demonstrate competency in oral and written communication as well as the ability to work in diverse teams. Engineering, maybe more than other occupations, need accurate and effective communication—I need to comprehend what you're saying and vice versa for the design we're both working on to operate.

The surprise aspect of communications is not that engineering schools have just begun to stress it (driven in some cases by ABET, but that there was ever a need to remind educators that engineers need to communicate! However, in a flat world, communication takes on a far larger meaning.

Not only are engineers routinely working on goods that will be manufactured in another nation and marketed to people of different cultures, but product engineering is increasingly being done by teams comprised of individuals from many countries and with varying cultural backgrounds. Such conversations obviously involve a high risk of misunderstanding and conflict.

Engineering schools can no longer approach preparing young engineers for employment in a flat world as an extracurricular activity, available exclusively to those with the time and finances to spend an extra semester overseas.

Every student must now cultivate the attitudes and abilities required to function worldwide, straight from the start of their working lives with expertise becoming a commodity, the future engineer must be capable of doing more than just doing technical duties. Engineers with the inspiration, vision, passion, and perseverance to revolutionize the way we live have always existed.

Those who have not, on the other hand, have previously been able to make a career performing ordinary technical duties. Young engineers of the future, on the other hand, must be exceptional. They will not be able to enjoy the convenience of well-paying occupations where regular duties are performed year after year.

The engineer of the future will be more responsible for developing new ideas and solutions and bringing them through to completion. Innovation has previously been acknowledged as one of the most essential determinants in both national and individual future success.

However, the engineering hurdles are far higher. Not only must the engineer be able to invent, but he or she must also be able to assist in making the concept a reality. Thus, the education of tomorrow's engineers must equip them to perceive new prospects as well as to marshal the resources needed to actualize their ideas.

2.8.1. Source of Human Asset—Pipeline Issues

Engineering courses have become a market, with students all over the world being able to access them over the internet. What distinguishes an engineering graduate from those from other countries? To remain competitive, we must produce inventive leaders for a technologically advanced society.

The new curriculum for the twenty-first century should prioritize innovation, creativity, and entrepreneurship, as well as the societal context of engineering. The relationship between the engineering profession and society requirements should be thoroughly described; this will motivate and attract students to the field.

The problem in the United States is that interest in engineering is diminishing, particularly among white males. Furthermore, when we look at the "production" of engineering graduates throughout the world, we observe that the United States lags behind several of the G7 countries.

What is most concerning is our kids' performance in the fundamental sciences in relation to many other nations. It is apparent that significant work is needed to rekindle interest in engineering and to convey that engineering is a social business.

One avenue to attract students and to address many of the needs described above is through internship programs such as the ones we have instituted at the Metal Processing Institute (MPI) at WPI. The underlying philosophy of the MPI Internship Program is to provide an educational experience, which is meaningful to the student, is contextual, and one that instills excellence.

The Internship Program bridges the gap between classroom learning and industrial experience. Unlike co-op programs, the internship program ensures that the industrial internship project is tied in with the academic plan of the student.

The MPI Internship provides a holistic and contextual educational experience — a new paradigm in graduate education. The internship concept is applicable to both undergraduates and graduate students.

2.8.2. Proactive Recommendations

Engineers solve problems, create things, and improve the quality of life on Earth. This has always been a constant; nevertheless, what has changed over time are our society's requirements and how engineers have responded to those needs. Engineers are credited with significant advancements and inventions to address the requirements of the Industrial Revolution in the late 1800s.

Engineers created things, built bridges, and developed mass manufacturing, transforming us from an agricultural to an industrial civilization. With developments in solid-state physics and our understanding of atomic structure in the 1900s, engineers learnt science and became scientists because they required the science foundation to tackle societal challenges.

This involves, among other things, the necessity for defense (a bomb, supersonic aircraft, armament), the discovery of the semiconductor, and the electronic materials revolution (Information age). Globalization and the "flattening of the World" are realities that are redefining the role of engineers and engineering as we approach the twenty-first century.

Engineers in the twenty-first century must be innovative and leaders in order to fulfill our society's requirements (global society). With 20% of the world's population living in absolute poverty, 18% lacking access to safe drinking water, 40% lacking access to sanitation, energy consumption increasing faster than population growth, and healthcare needs and expectations increasing faster than the cost of health care delivery, there is no doubt that the engineer of the twenty-first century must be a social scientist as well as an enterprising leader.

There are particular steps that we as a community should explore moving ahead; the underlying theme is that we need to improve the image of engineering and feed the "innovation engine."

2.8.3. Change of Image

Currently, the public's perception of engineers and engineering does not represent reality. Many of our top industrialists and successful chief executive officers (CEOs) are engineers, and many surgeons and physicians have a first degree in engineering. We have bankers and financial tycoons who have engineering degrees. There are no restrictions.

Engineering's image must be altered to reflect the limitless prospects and lifestyles that await its graduates. Back in the early 1900s, engineering instructors did not pay attention to management difficulties, allowing management to virtually depart the engineering curriculum; we have witnessed the emergence of Master of Business Administration (MBAs), particularly after WWII. This was a blunder.

This was the Polytéchnicien's opinion in 1794. Perhaps we should return to this picture and educate our youth about the leadership prospects that engineering provides. Furthermore, we want a consistent message. Currently, the messaging about engineering as a professional choice is fractured.

Civil engineers (ASCE), mechanical engineers (ASME), metallurgists, and materials scientists and engineers (ASM, TMS), electrical engineers, and chemical engineers (AICHE) all have different messages. The messages differ and should be the same — we need a consistent message.

2.9. SOCIETAL ISSUES AND ENGINEERING PROFESSION AS AN ENABLING PROFESSION

Our global population will expand to around 9.5 billion people (from 6.5 billion) and over twenty-first century, with emerging countries accounting for the majority of this expansion. The societal requirements for energy resources, transportation, housing, packaging materials/recycling, biomaterials, and health will only increase.

The obstacles we confront in achieving global sustainability are enormous. This is precisely why engineering should be so appealing to the next generation; we must demonstrate that engineering is an enabling profession.

The case for engineering as an enabling profession for global sustainable development is compelling; nevertheless, this relationship is rarely articulated explicitly.

2.10. REDEFINING THE EVER-EVOLVING ENGINEERING PROFESSION

Engineering in the twenty-first century is at the epicenter of a knowledge boom. New and innovative discoveries in science, engineering, medicine, mathematics, and the social sciences have changed how we see and interact with the world around us, as well as led to the erasure of academic disciplinary boundaries.

Engineering is the catalyst that brings disciplines together and improves the progress made possible by such partnerships. This modern approach necessitates the capacity to transcend one's own background and connect with specialists from other and divergent fields.

Engineers aid fields such as medicine, veterinary science, geology, atmospheric science, chemistry, biology, and other life sciences progress in a synergistic manner. Technical experts working in a range of businesses must keep up with improvements in their disciplines, acquire new techniques to dealing with classic difficulties, and recognize new concerns that occur over time and how to deal with them.

It was simple to follow technical publications 300 years ago since knowledge expanded at a slower rate and there were few journals. Scientific journals began increasing rapidly and doubling in number every 15 years as publishing expanded. In 1987, it was estimated that there were around 5000 scientific periodicals in the United States.

According to some current estimates, the present global paper production "amounts to more than two million technical papers every day, or a daily output that would fill seven sets of Encyclopedia Britannica." Current techniques employ computer abstracting and computer searches of vast national and international data bases, which contribute to the ever-increasing fruitful explosion of information.

Even in a very restricted field of specialty, no single technical specialist can know or access all necessary knowledge. It has been reported that engineers as professionals use published technical literature much less than scientists and other professionals because major indexing is done by subject, which lends itself well to use by scientists but is not very helpful to practicing engineers dealing with complex design situations in changing conditions.

Professional engineers advise engineers to create their own data bases addressing difficulties specific to their job. Experienced engineers advise keeping files containing articles of important information. Engineers should also engage with vendors who have their own relevant information while thoroughly analyzing what was acquired.

2.11. THE DECREASED DURABILITY OF THE INITIAL ENGINEERING EDUCATION

The wide basis of an engineer's background is intended to allow for the acquisition of skills as one practices the profession. However, sooner or later, one must focus on problem-solving abilities inherent in the job setting. As time passes, some essential information is forgotten, and abilities become rustier and less efficient.

And if the engineer is involved in management matters, his or her technical competency suffers even more. As a result, the necessity to refresh existing abilities and learn new ones becomes more pressing and a question of professional survival. This need is becoming more important as the labor market grows more competitive.

Many engineers maintain their currency by participating in a series of training programs that lead to a postgraduate degree or certificate, or just to improve their competence in a certain field. These might be classes offered by universities or other educational institutions.

Some courses may be provided by professional societies or through long-distance learning programs, but others may be provided at the employer's site and are often designed to boost the productivity and sophistication of the employee's contribution in the workplace.

Studies of professionals inside bigger industrial companies revealed that having a graduate degree increased the career progression of engineering experts. Furthermore, their levels of performance were superior to individuals with merely a BS degree. It was also found that 10 years after the performance of BS holders began to deteriorate, the contributions of postgraduate degree holders were maintained.

This clearly implies that "a substantial dose of graduate courses can put obsolescence back by 10 years.." Courses held on the premises of the business are generally appealing to senior professionals who are more at ease in surroundings without official tests and grading and who feel less stressed than in competitive environments of institutions of higher learning.

Personal initiative is less necessary in major metropolitan regions, where a greater number of engineering professionals desire to advance in their careers. Of course, these are more important in small towns and cities with one or two industrial enterprises.

The emphasis on distance learning options appears to be the logical approach to seek out extra educational instruction and skill upgrading. Interaction with colleagues is a strong form of knowledge expansion. Meeting specialists who are fully involved in their practice has a powerful motivating effect, especially when the on-the-job experience is at the leading edge of the discipline. When this is combined with on-the-job mentorship, learning is truly increased. When an industrial outfit's management is supportive of the learning experience and provides adequate support to advance professional skills and challenging job assignments by allowing attendance at professional meetings and conventions, short courses, industrial fairs and exhibitions, this results in an improved intellectual climate and a vigorously creative and progressive workplace.

2.12. DETERMINING WHETHER ENGINEERING IS A PROFESSION HERE OR THERE

The usage of the word "profession" in anything resembling the type of honest work outlined here appears to have originated only in the last hundred years or so in English-speaking nations, and has expanded only in the last 50.

There is, I believe, little reason to dispute that "profession" (in the sense defined here) is an English innovation, much like the railroad engine and parliamentary democracy, and that it has spread to much of the rest of the globe. Every new thing has to start somewhere.

Nonetheless, several non-English-speaking nations that do not have a (formal) code of professional ethics or their own name for profession appear to have entities that are essentially similar to professions in English-speaking countries"). As a result, having a formal rule or needing it to apply to anything termed a "profession" appears unduly Anglo-centric, as well as deciding by definition what might otherwise be an intriguing empirical subject.

It is also critical that the Socratic definition presented here does not need a profession to have a formal code of ethics (or to be termed "a profession"), but rather teaches us on how to assess whether a certain employment is structured in a certain manner through empirical investigation. What it instructs us to search for is the previously mentioned tripartite relationship between occupation and morality.

According to the Socratic definition, it is this complicated link that (and no other), separates profession from many other types of social organization that are essentially comparable, such as labor unions, learned organizations, and licensed trades.

In many nations where formal codes of professional ethics do not exist, technical standards encompass the same criteria that a code of ethics would in England, Australia, or the United States, but implicit in specifics rather than stated in the broader language of a code of ethics. In certain nations, the code of ethics may be both written and "unwritten" (i.e., not institutionalized as a "code of ethics").

Whether or if the technical standards of engineers in a given nation function as an implicit code of ethics is determined by the attitude that engineers in that country have toward those standards (provided the standards exist). must be ethically acceptable and created to fulfill the same ideal as other engineers).

If engineers in a given nation see local technical standards as (mainly) foreign impositions, the standards are considered legislation rather than a (implicit) code of ethics (whatever their content).

If, on the other hand, each engineer (or, at least, the majority of them) regards them as standards that every other engineer there should follow even if it means doing the same, that is, as part of a cooperative practice, then (all else being equal), the standards do constitute a code of ethics (even if an unusually detailed one and even if enacted into law)—and a profession of engineering exists there.

For more than a decade, I have conducted such informal empirical research, primarily by asking questions of engineers or engineering professors I encounter on my travels. I have the feeling that other nations, such as the Netherlands, definitely have an engineering profession, even if its engineers lack a codified code of ethics or a word for their job that is not derived from English.

I also believe that some countries may be lacking in the engineering profession. For example, the French engineers I spoke with appeared to regard themselves as government agents (even if working for a private employer). They worked for "the state" (l'état), rather than a more recognizable moral ideal (such as "public good").

They saw themselves as obligated by law and morals, but not by a professional ethics code (as I understand that term). Indeed, they originally mistook "profession" for "occupation" (even when speaking English) and struggled to comprehend what I meant by "professional ethics." They mistook my intention for the application of moral ideas (what philosophers teach in "Ethics" courses) to engineering.

Despite these conceptual issues, their solutions to real engineering ethics concerns, including the justifications they provided, appeared to follow what engineers in other countries would say. As a result, I see France as a "interesting instance" rather than a clear example of a country with engineers but no engineering profession.

My view of Japanese engineers who are quite competent in English is that they are more like the Dutch than the French, although the majority of them owe their proficiency in English to having had part of their technical education in the United States. They cannot be considered a representative sample of average Japanese engineers.

We won't know how far the engineering profession goes in Japan or elsewhere until we (well, social scientists) travel to all those sites and ask engineers questions about their job and their relationships with other engineers. My concern about previous research is that researchers asked the incorrect questions (questions recommended by one or more sociological conceptions) and so learned a lot about specific jobs but essentially nothing about professions as a whole.

Understanding engineering as a profession has several implications for education and research. So, if engineering is a profession everywhere, all engineers are engineers (and only engineers) belong to a single community, engineering—whether they also belong to other communities, such as a nation, linguistic group, firm, industry, or job type ("technologists").

To understand engineers as engineers, we must first comprehend their vocation (as well as their function, discipline, and occupation). If we are to teach engineering ethics, we must consider not only the substance of their code of ethics, but also the unique reason a professional has to follow it ("Don't cheat").

2.13. CONCLUSION

According to the most recent US Census Bureau estimates and predictions, the world population will reach 9.3 billion people by 2050. Various societal and engineering profession issues have also been discussed in the chapter .

More is required of engineers as a result of global socioeconomic demographic trends, environmental concerns, and the earth's finite resources. It has taken us from the dawn of time to achieve a population of one billion. People, regardless of where they reside, require the essential necessities: food, clothes, shelter, and water for drinking or cleansing.

At the turn of the twentieth century, there were around six billion of us on the planet. Future engineers are expected to produce goods and services that improve living standards and health care while simultaneously addressing severe environmental and sustainability challenges. We also want to feel secure, to be able to rest and be entertained.

REFERENCES

1. Saeed Moaveni, (2011). *Engineering Fundamentals*. [e-Book] https://www.hzu.edu.in. Available at: https://www.hzu.edu.in/engineering/Engineering%20Fundamentals.pdf (accessed on 1 July 2022).
2. Apelian, D., (2007). *The Engineering Profession in the 21st Century–Educational Needs and Societal Challenges Facing the Profession*. [e-Book] https://www.tms.org/. Available at: https://www.tms.org/Communities/FTAttachments/Apelian-Hoyt.pdf (accessed on 1 July 2022).
3. Davis, M., (2008). Is engineering a profession everywhere?. *Philosophia, 37*(2), 211–225. [online] Available at: https://www.researchgate.net/publication/226215236_Is_Engineering_a_Profession_Everywhere/link/53e227e90cf2235f352c1357/download (accessed on 1 July 2022).
4. Dias, P., (2019). Introduction: From engineering to philosophy. *Philosophy for Engineering,* 1–7. [online] Available at: https://link.springer.com/chapter/10.1007/978-981-15-1271-1_1 (accessed on 1 July 2022).
5. Kalghatgi, G., Agarwal, A., Senecal, K., & Leach, F., (2021). Introduction to engines and fuels for future transport. *Energy, Environment, and Sustainability*, 1–5. [online] Available at: https://link.springer.com/chapter/10.1007/978-981-16-8717-4_1 (accessed on 1 July 2022).
6. Kreiner, J., (2006). *Combating Obsolence–Redefining the Ever-Evolving Engineering Profession*. [e-Book] https://www.mas.bg.ac.rs. Available at: https://www.mas.bg.ac.rs/_media/istrazivanje/fme/vol34/4/8._kreiner_227-230.pdf (accessed on 1 July 2022).

CHAPTER 3

BASICS OF MECHANICAL ENGINEERING

CONTENTS

Mechanical engineering plays a significant role in improving global safety, economic vitality, enjoyment, and general quality of life. Mechanical engineers study the fundamentals of force, energy, and motion. Mechanical engineering is a broad field that emerges from the necessity to develop and build anything from small individual parts and devices (e.g., microscale sensors and inkjet printer nozzles) to enormous systems (e.g., spacecraft, and machine tools).

A mechanical engineer's job is to take a product from concept to market. A wide range of talents are required to do this. Mechanical engineering is likely the largest and most diversified engineering subject since these abilities are necessary for almost everything that is created.

3.1. INTRODUCTION

Mechanical engineers are essential in industries such as automotive (from the car chassis to every subsystem — engine, transmission, sensors); aerospace (airplanes, aircraft engines, control systems for airplanes and spacecraft); biotechnology (implants, prosthetic devices, fluidic systems for pharmaceutical industries); computers and electronics (disc drives, printers, cooling systems, semiconductor tools); and microelectromechanical systems (MEMS) (sensors, actuators) (machining, machine tools, prototyping, micro fabrication).

Energy conversion (gas turbines, wind turbines, solar energy, and fuel cells); environmental control (HVAC, air conditioning, refrigeration, and compressors); automation (robots, data, and image acquisition, identification, and control); and manufacturing (machining, machine tools, prototyping, micro fabrication).

3.2. WHAT IS MECHANICAL ENGINEERING?

A branch of engineering that deals with the development and use of heat and mechanical power, as well as the design, manufacture, and usage of machines. A machine is an apparatus that uses or applies mechanical power and has numerous pieces, each with a specific purpose and working together to execute a certain activity (Figure 3.1).

Figure 3.1. Mechanical engineering gear.

Source: Image by pixabay.

Mechanical engineering, as the name implies, is concerned with the mechanics of mechanical system operation. This engineering discipline encompasses the design, analysis, testing, manufacture, and maintenance of mechanical systems.

A mechanical engineer could create a component, a machine, a system, or a process. Mechanical engineers will assess their design using motion, energy, and force principles to guarantee that the product performs safely, efficiently, and reliably, and that it can be built at a competitive cost.

3.3. BASIC CONCEPTS

3.3.1. Force

A force is a fundamental notion in physics that may be defined as any influence that tends to modify the motion of an item. A force is defined as the push or pull exerted on an object as a result of its contact with another object. When two objects interact, a force is exerted on each of the items. When the interaction ends, the force is no longer felt by the two objects.

Forces arise exclusively as a result of contact. The cosmos is governed by four fundamental forces: gravity, nuclear weak force, electromagnetic force, and electromagnetism. The nuclear strong forces in ascending strength order in mechanics, forces are considered to have been the causes of linear motion, whereas torques are thought to be the causes of rotating motion.

Newton's Laws describe the operation of forces in creating motion. Force is a quantity that would be measured using the Newton, the standard metric unit. A Newton is symbolized by a "N." "10.0 N" indicates 10 Newtons of force. One Newton is the amount of force necessary to accelerate a 1 kilograms mass by 1 m/s^2.

A force is a vector quantity, which means that it has both magnitude and direction. To adequately explain a force operating on an item, you must specify both its magnitude (size or numerical value) and its direction. As a result, 10 Newtons does not adequately describe the force operating on an item (Figure 3.2).

Figure 3.2. Mechanical engineering.

Source: Image by Wikimedia commons.

In comparison, 10 Newtons downwards is a comprehensive description of the force exerted on an item; it specifies both the magnitude (10 Newtons) and the direction (downwards). Torque is a type of force that twists an axle in a specific direction. It is also known as a rotational force.

Torque may be generated by pressing on a rod or lever that turns an axle. Similarly, a torque on an axle might result in a linear force at a distance from the axle's center. Torque is defined as force multiplied by moment arm. A torque may be created on an axle by pushing on a rod that turns it. A torque on an axle, for example, can result in a linear force at a radius from the center.

3.3.2. Work

Work is defined as any action that involves a force and movement in the direction of that force. When a force operates on an object in the direction of motion or has a component in the direction of motion, work is done on it. To do work on an item, a force must be applied to it, and the object must move in the direction of the force. Joules are units of measurement for work (J).

3.3.3. Energy

Energy is the ability to do tasks. To do work, you would have to have energy — it is the "money" for just doing labor. During the process of doing work, the thing doing the job exchanges energy with the object being worked on. When a thing is worked on, it gains energy. Mechanical energy is the energy gained by the things on which labor is performed.

Mechanical energy is the energy that an item possesses as a result of its motion or location. Mechanical energy can be kinetic energy (motion energy) or potential energy (stored energy of position). Objects contain mechanical energy if they are in motion and/or at some point in relation to a potential energy position of zero.

3.3.4. Power

The rate at which work is completed is referred to as power. It's the work-to-time ratio. It is calculated mathematically using the following equation. The Watt is the standard metric unit of power. A unit of power is equivalent to a unit of labor divided by a unit of time, as represented by the power equation.

As a result, a Watt equals a Joule/second. The term "horsepower" is occasionally used to denote the power provided by a machine for historical reasons. One horsepower is about equal to 750 Watts. The majority of machines are created and manufactured to perform tasks on things.

A power rating is commonly used to characterize all machinery. The power rating of a machine reflects how quickly it can do work on other items. As a result, a machine's power is defined as its work/time ratio. The power rating refers to how quickly the engine can accelerate the vehicle.

3.3.5. Friction

Friction is the force caused when two surfaces move or attempt to move across one other. Friction is always an opposing force to the mobility or attempted motion of one surface over the other. Friction is determined by the texture of both surfaces as well as the amount of force pressing the two surfaces together.

In a machine, friction decreases the output to input ratio. A car, for example, spends one-quarter of its energy decreasing friction. However, friction in the wheels allows the automobile to stay on the road, and friction in the clutch allows the vehicle to move at all.

Friction is one of the most important physical phenomena, affecting everything from matches to motors to molecular structures. Friction has both advantages and downsides. Friction, which is a resisting force that slows or stops motion, is required in many applications to avoid slipping or sliding.

However, it can be inconvenient since it might impede mobility and necessitate the expenditure of energy. To get precisely the right amount of friction, a good compromise is required. When two components come into contact, the force of friction acts to oppose motion.

There are small bumps and ridges on the surfaces of all substances (Figure 1.2). When two components move past each other, such minuscule peaks and valleys interact.

3.4. COMPUTER-AIDED DESIGN

Throughout the life cycle of engineering goods and manufacturing processes, computers play a major, and frequently dominant, role. Their job is critical as worldwide competitive pressures demand advances in product performance and quality while reducing product design, development, and production timelines significantly (Figure 3.3).

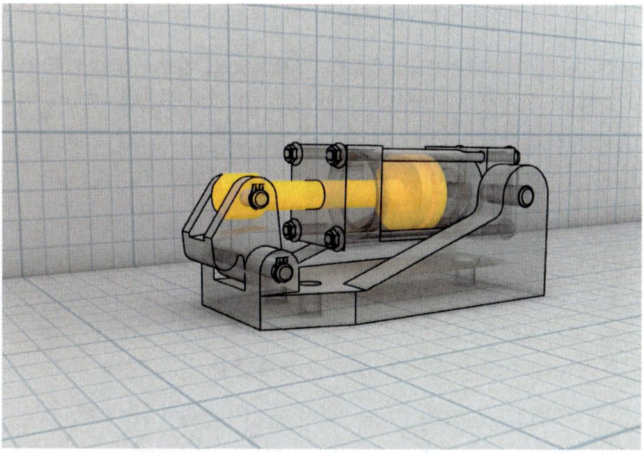

Figure 3.3. CAD design – mechanical engineering.

Source: Image by Maxpixel.

Using computers, design engineers significantly increase their job productivity. For example, using proper simulation software, the performance of a product or process may be tested prior to building a prototype. Computer-aided design (CAD) makes use of the computer's mathematical and graphic processing capacity to assist engineers in the development, modification, analysis, and display of designs.

Many causes have contributed to CAD technology being a critical tool in engineering for applications such as shipbuilding, automobile, aerospace, medical, industrial, and architectural design, for example, the computer's speed in solving difficult equations and handling technical databases.

CAD was formerly assumed to be merely computer-aided drafting, and its usage as an electronic drawing board remains a strong tool in and of itself. A CAD system may also be used to do geometric modeling, engineering analysis, simulation, and design information sharing.

However, a CAD system's capabilities are expanding much beyond its capacity to represent and modify visuals. As part of collaborative product design (CPD), sustainability effect analysis, product life-cycle management, and product data management (PDM), the CAD system is being incorporated into the whole product life.

3.4.1. Historical Perspective on CAD

In many aspects, CAD is founded on graphical data representation. In the 1950s, computer graphics were employed in the Semi-Automatic Ground Environment (SAGE) Air Defense Command and Control System. SAGE turned radar data into computer-generated pictures that were shown on a cathode ray tube (CRT).

It also employed a light pen as an input device to choose information straight from the CRT screen. Another notable innovation in computer graphics technology happened in 1963 when Ivan Sutherland presented the SKETCHPAD system in his PhD thesis at MIT.

The SKETCHPAD system was powered by a Lincoln TX-2 computer. SKETCHPAD is a graphical user interface that allows a design to be entered into a computer through a light pen on a CRT monitor. Using the light pen, pictures may be generated and modified in SKETCHPAD.

SKETCHPAD allowed for on-screen graphic modifications such as translation, rotation, and scaling. Sutherland's method has resulted in computer programs known as interactive computer graphics (ICG), which serve as the foundation for CAD design processes. SKETCHPAD's graphical capabilities demonstrated the possibilities for digital sketching in design.

Sutherland continued his research on helmet-mounted displays (HMDs), the predecessor to virtual reality head displays, as a professor of electrical engineering (EE) at the University of Utah. The field of computer graphics, as we know it today, sprang from the numerous new ideas and breakthroughs generated by the scholars who established the University as a center for this type of study.

Sutherland co-founded Evans and Sutherland in 1968 with Dave Evans, the founder of the University's Computer Science Department, and went on to pioneer computer modeling methods and software. Sutherland was the chairman of the Computer Science Department at the California Institute of Technology from 1976 until 1980 (Figure 3.4).

Figure 3.4. Mechanical engineering shop.

Source: Image by Wikimedia commons.

While there, he assisted in the introduction of integrated circuit (IC) design to academics. They co-developed the science of merging computational mathematics with the physics of actual transistors with Professor Carver Mead and genuine wires, and went on to establish IC design as a legitimate topic of academic study.

Sutherland left Caltech in 1980 and founded Sutherland, Sproull, and Associates. Sun Labs purchased the company in 1990, laying the groundwork for Sun Microsystems Laboratories. Because of the high cost of computer hardware in the 1960s, ICG systems were only used by major firms such as those in the automobile and aerospace sectors, who could justify the initial investment.

Computers got more powerful as computer technology advanced, with faster processors and better data storage capacities. As computer prices fell, systems became more affordable to small businesses, allowing entrepreneurs to experiment with CAD tools and technologies.

Advances in Web-based technologies and standards, the usage of mobile computing platforms and devices, cloud-based storage, software as a service (SaaS), and functional integration into enterprise-wide systems have all contributed to the growing influence of computer-aided design in recent years.

Furthermore, the proliferation of CAD systems that operate on a wide range of platforms has facilitated worldwide cooperation as well as concurrent design and production processes. Many consider CAD to be a crucial business tool for any engineering, design, or architectural firm.

3.4.2. Design Process

Before delving into computer-aided design, it's vital to comprehend the design process in general. What are the circumstances that lead to the start of a design project? What steps does an engineer take while designing something? How does one come to the decision that the design is finished?

We address these concerns by categorizing the process into six discrete stages:

- Customer or sales field input and need perception;
- Problem identification;
- Initial design;
- Evaluation and optimization;
- Evaluation and testing; and
- Design and specification final.

A need is often regarded in one of two ways. Someone must notice a flaw in a current design or a consumer-driven opportunity in the marketplace for a new product based on sales field reports or customer feedback. In either situation, there is a requirement that may be met by changing an existing design or establishing an altogether new design.

Because the need for change may be revealed only by subtle conditions, such as noise, growing sustainability concerns, marginal performance characteristics, or departures from quality standards, the design engineer who detects the requirement has made a first step toward resolving the problem.

This stage initiates processes that may assist others to understand the need more clearly and maybe join the solution process. Once the choice has been taken to take corrective action in response to the requirement, the problem must be characterized as a specific problem to be solved, with all important characteristics identified (Figure 3.5).

Figure 3.5. Mechanical engineering design – CAD.

Source: Image by Flickr.

These restrictions frequently include pricing constraints, quality requirements, size, and weight constraints, and functional constraints. Specifications are frequently established by the capabilities of the production process. Anything that will impact the engineer's choice of design characteristics must be mentioned in the problem definition.

Careful planning at this step might result in fewer iterations in later phases of design. After completely defining the problem in this manner, the designer proceeds to the preliminary design stage, when expertise and creativity may be utilized to develop an early design idea at this point, collaboration may help the design be more successful and effective.

This design is then subjected to several types of examination, which may identify particular flaws in the original design. The designer then uses the analytical data to an iteration of the preliminary design stage. These iterations may be repeated numerous times through early design and analysis cycles until the design is optimized.

The design is subsequently tested and assessed using the parameters specified in the problem specification. A scale prototype is frequently built to conduct more analysis and testing to evaluate operational performance, quality, dependability, and other criteria. If a design problem is discovered during this stage, the design will be abandoned.

The design is returned to the preliminary design/analysis stages for redesign, and the process is repeated until the design has passed the testing step and is ready for presentation. The last stage of the design process is the final design and specification. It is critical to communicate the design to others in such a way that its fabrication and marketing are viewed as critical to the firm.

When the design is totally accepted, comprehensive engineering drawings are created, replete with requirements for components, subassemblies, tools, and fixtures needed to construct the product, as well as the related production costs. These can then be sent physically or digitally to the various manufacturing divisions utilizing the CAD data.

3.5. CAD SOFTWARE

Graphics software uses the CPU and its peripheral I/O devices to produce a design and render it on screen, paper, or other output devices such as rapid prototyping. Analysis software makes advantage of the CPU's computing speed to perform dimensional modeling and different analytical procedures such as interference checks on the recorded design data.

CAD software used to be confined to graphics software (2D and 3D design) and analytical applications such as FEA, property analysis, kinetic analysis, and rapid prototyping. PLM, CPD, and PDM are examples of CAD software systems.

Modern CAD software is frequently supplied as "packages" that include all of the programs required for CAD applications. Modules and add-on packages can be purchased If the demand arises, the CAD software vendors.

3.6. 2D GRAPHICS SOFTWARE

Traditional drafting consisted of producing 2D technical drawings that were used in the synthesis stage of the overall design process. However, modern computer graphics software, like that used in CAD systems, allows designs to be portrayed pictorially on the screen, allowing the human mind to build perspective, providing the illusion of three dimensions on a two-dimensional screen.

Regardless of the design representation, drafting is just taking the conceptual answer to a previously identified and specified problem and showing it pictorially. As previously stated, one of the benefits of computer-aided design is the "electronic drawing-board" capability. The drawing board made available by CAD systems is the consequence of supporting graphics software.

That program makes graphical depiction of a design on screen easier by transforming graphical input into Cartesian coordinates along the x, y, and occasionally z axes. Geometric forms, for example, are frequently written directly into software for simpler geometric representation.

The coordinates of the user-created lines and objects may then be arranged into a matrix and manipulated using matrix multiplication for operations like scaling, rotation, and translation. The final points, lines, and forms are being sent to graphics software and, eventually, to the display screen for simpler design modifications.

Because the entire procedure may be completed in a matter of nanoseconds, the user gets the findings virtually instantly. While matrix mathematics provides the foundation for the movement and manipulation of a drawing, almost all CAD software is focused on easing the drafting process since generating the drawing line by line and shape by shape is a time-consuming and monotonous procedure in and of itself.

3.7. 3D GRAPHICS SOFTWARE

3.7.1. Wire Frame Modeling

A wire frame model is a 3D computer graphics depiction of an item in which all surfaces are plainly defined, including internal components and sections of the object that have been ordinarily concealed from view. A wire frame model enables for simple understanding of the object's underlying design.

However, it is the least practical approach of presenting a 3D object. Wire frame models are commonly used in surveying, hydrology, geology, and mining.

3.7.2. Surface Modeling

Surface modeling (also referred as free-form surface modeling) is a CAD technique for describing the exterior surface, or skin, of a 3D object which does not have regular radial dimensions. NURBS mathematics is frequently used to describe surfaces in surface modeling.

The mathematical characteristics, shape, and smoothness of transition of free-form surfaces are determined by their poles, degree, and patches (segments with spline curves). This methodology is more complicated than wire frame modeling but less complicated than solid modeling. Applications include developing turbine blades and automotive bodywork.

3.7.3. Solid Modeling

Three-dimensional geometric or solid-modeling capabilities follow the same fundamental notion as shown above, but with a few additional concerns. First, there are numerous techniques to constructing a three-dimensional design.

Second, several operators in solid-modeling software may be involved in the construction of the 3D geometry. There are several ways in CAD solid-modeling software that determine how the user generates the model.

Various functional methods to solid modeling have been developed since the introduction of solid-modeling capabilities into the CAD mainstream. Dimension-driven solid-modeling features, such as variational design, parametric design, and feature-based modeling, are now supported by several CAD software systems.

3.7.4. Dimension-Driven Design

Dimension-driven design refers to a system wherein the model is defined as a series of equations that must be solved sequentially. These equations allow the designer to establish limitations, such as requiring one plane to always be parallel to another.

If the first plane's orientation is modified, the angle of the second plane will also change to preserve the parallel connection. The term comes from the fact that the equations used to identify plane positions frequently specify the distances between data points.

3.7.5. Variational Modeling

The variational modeling approach defines the design as a sketch, which can then be easily translated to a 3D mathematical model with fixed dimensions. If the designer changes the design, the model must be recalculated entirely. Because it takes the dimension-driven approach of handling equations sequentially and makes it no sequential, this approach is quite flexible.

Dimensions may then be changed in any order, making it an ideal for usage early in the design phase when the geometry of the design may alter considerably. By avoiding the need to calculate any extraneous equations, variational modeling reduces computing time (and hence increases program run speed).

Creating 2D profiles of the design that may depict cnd vicws and cross sections is what variational sketching entails. Using this method, the designer often concentrates on generating the ideal shape while paying minimal attention to dimensional details. Once the design form has been developed, the design may be scaled to the necessary dimensions using a separate dimensioning function.

3.7.6. Parametric Modeling

Engineering equations between sets of parameters, such as size parameters and geometric parameters, are solved using metric modeling. Size parameters are dimensions such as a hole's diameter and depth. Geometric parameters, such as tangential, perpendicular, or concentric connections, are limitations.

Approaches to metric modeling keep a record of operations performed on the design so that relationships between design elements can be inferred and incorporated into later changes in the design, allowing the change to be made with a certain degree of acquired knowledge about the relationships between parts and design elements.

3.7.7. Feature-Based Modeling

Feature-based modeling allows the designer to create solid models from industrial standard geometric elements such as holes, slots, shells, grooves, windows, and doors. A "thru-hole" feature, for example, can be used to specify a hole.

When this feature is activated, the hole will always remain open on all sides, regardless of the thickness of the material through which the hole travels. In variational modeling, however, if a hole is made in a plane with a predetermined thickness and that thickness is raised, the hole becomes a blind hole until the designer adjusts the size of the hole to offer an opening at both ends.

The key advantage of feature-based modeling is that design intent is maintained independent of dimensional changes in the design. Another key benefit of a feature-based approach is the flexibility to update multiple design aspects in response to a change in a specific portion.

For example, if the threading of a bolt is changed, the threading of the related nut is immediately updated, and if that bolt design is used several times in the design, all bolts and nuts may be modified in a single step. In certain feature-based CAD systems, a knowledge base and inference engine make feature-based modeling more intelligent.

3.8. GRAPHICAL REPRESENTATION OF IMAGE DATA

Wire frame representations in 2D and 3D can be misleading and difficult to comprehend. Because mechanical and other engineering designs frequently contain 3D components and systems, 3D representation-capable CAD systems have fast become the most common in engineering design.

To build a 2D view from a 3D model, the CAD program must be supplied information specifying the user's viewpoint. The computer may then use this information to calculate angles of view and identify which surfaces of the design are visible from the particular viewpoint. Surfaces nearest to the observer are generally used by the program to block out surfaces that would otherwise be obscured from view (Figure 3.6).

Figure 3.6. Male mechanical engineer solders parts for prosthetic limbs.

Source: Image by Flickr.

Then, using this strategy and working in the opposite direction of the spectator, Which surfaces are visible are determined by the program. The following stage calculates the virtual distance between the viewer and the model, enabling areas outside the screen's bounds to be eliminated from consideration.

The colors shown on each surface must then be decided by integrating user selections for light source and surface color. The simulated light source may play a significant role in portraying the picture accurately by affecting the color values used for the design and defining reflection and shadow locations.

High-end CAD software can now replicate a number of light sources, such as spot lighting and sunshine, either directly or through an aperture such as a door or window. Surface textures can even be selected and shown in some systems. Following these judgments, the program computes the color and value for each pixel of the display.

Because all these computations are computationally costly, the hardware employed is frequently equally as critical as the software when using solid-modeling applications with advanced surface representation features. The minimal and optimal hardware system requirements for CAD software are specified by the program providers.

3.9. ANALYSIS SOFTWARE

The simulation of a developed device's performance is a crucial aspect of the design process. A fastener, for example, may be designed to perform under specific static or dynamic loads, or the temperature distribution in a CPU chip may need to be analyzed to predict heat transfer behavior and potential thermal stress.

Alternatively, turbulent flow across a turbine blade regulates cooling but may cause vibration, which must be addressed. Whatever the equipment being developed, there are several variables on its performance that may be identified using CAD analysis tools. FEA may be used to calculate the load categories indicated above.

The analysis breaks down a given domain into smaller, identifiable essential portions known as elements. The needed mathematics is then used to analyze each constituent. Finally, the overall solution to the issue is established by aggregating the individual solutions of the constituents.

Thus, complex issues can be solved by breaking them down into smaller, simpler problems to which approximate solutions can be applied. General-purpose FEA software tools have been generalized to the point that users do not need to be experts in FEA.

A FEM may be considered of as a collection of solid bricks (elements) put together. The finite-element library contains several types of elements, which are listed below. NASTRAN by NEi Software (previously Noran Engineering, Inc.) and ANSYS by ANSYS Inc. are two well-known general-purpose FEA software.

3.10. CAD STANDARDS AND TRANSLATORS

Four major formats permit data sharing in order for CAD software to function across systems from diverse suppliers:

3.10.1. Initial Graphics Exchange Specification (IGES)

IGES is an ANSI standard for digital representation and information transmission between CAD/CAM systems. IGES may be used to transform 2D geometry and 3D CSG. IGES versions have support for B-rep solid modeling.

IGES file parsing and formatting, generic entity manipulation procedures, common math utilities for matrix, vector, and other applications, and a powerful collection of geometry conversion methods and linear approximation capabilities are all accessible in the IGES library.

3.10.2. STEP (Standard For the Exchange of Products: ISO 10303)

STEP is an international standard collection. It provides a single natural format for CAD data that may be used throughout a product's life cycle. STEP provides capabilities and benefits that IGES does not. The user can utilize an IGES standard to obtain all of the necessary data in a single document.

STEP can also be used to exchange B-rep solids across CAD programs. STEP varies from IGES in how data is defined. The user in IGES gets out the specification, reads it and then implements what it says.

The implementor in STEP takes the definition and runs it through a particular compiler, which produces the code. This procedure ensures that there is no ambiguity in data interpretation across implementors.

3.10.3. Drawing Exchange Format (DXF)

DXF is the de facto standard for sharing CAD/CAM data on a PC-based system, established by Autodesk, Inc. for AutoCAD software. Only 2D drawing data in ASCII or binary format may be converted into DXF.

DWG (the name of Autodesk's proprietary file format) is a development of the DXF format, allowing the storage of both 2D and 3D data as well as related parameters. DWG files are natively supported by AutoCAD and other CAD applications.

3.10.4. American Committee for Interoperable Systems (ACIS)

The ACIS modeling kernel is a collection of software algorithms used in the development of solid-modeling tools. Spatial Corp. (previously Spatial Technology Corp.) licenses ACIS routines to software developers to make the work of creating new solid modelers easier.

The main advantage of this technique is that models built with ACIS-based software should be compatible with other brands of ACIS-based modelers. This eliminates the requirement to employ IGES translators to transport model data across programs.

ACIS-based products for CAD/CAM and FEA software packages are now commercially available. ACIS output document contains the suffix *.SAT or *.SAB.

3.11. APPLICATIONS OF CAD

3.11.1. Optimization Applications

Engineers want rapid, dependable tools as designs get increasingly complicated. Finite-element analysis has evolved into the primary method for identifying and solving design issues. Although the increased design efficiency given by CAD has been supplemented by the use of FEMs to analysis, engineers still frequently utilize a trial-and-error technique to solve errors highlighted by FEA.

This technique invariably increases the time and effort required for design since it requires more time for contact with the computer, and solution options are frequently restricted by the designer's own experiences. Design optimization tries to reduce most of this excess work by employing a rational mathematical technique to ease difficult design adjustment.

The goal of optimization is to decrease or maximize a characteristic, such as weight or physical size, that is constrained by one or more factors. The technique utilized to optimize a design is determined by its size, form, or both. Typically, optimizing the size of a design is simpler than optimizing the form.

The geometry of a plate does not change considerably when its thickness is optimized. Optimizing a design parameter, such as the radius of a hole, on the other hand, changes the geometry during form optimization. Optimization methods were difficult to adopt in the engineering setting since the process was somewhat intellectual in character and was not seen as easily transferable to design processes.

Optimization techniques, on the other hand, may be easily understood and utilized in the design process when regarded as a component of the process itself. Up to a point, iterations of the design method proceed as they generally do in design. The designer then puts the optimization software into action.

The optimization's objectives and limitations must first be specified. The optimization tool then examines the design in relation to the objectives and limitations and makes design improvements automatically. Because the process is automated, engineers should be able to monitor the design's progress during optimization, stop the program if required, and restart.

The strength of optimization programs is primarily determined by the capabilities of the design tools employed early in the process. Automatic and parametric meshing skills are required for two-and three-dimensional applications. Other applications necessitate linear static, natural frequencies, mode shapes, linearized buckling, and steady-state analysis.

Because the design geometry and mesh might vary throughout optimization cycles, the optimization algorithm must contain error estimation and adaptive control. Furthermore, when distinct pieces are to be combined and examined as a whole, the software frequently finds it useful to link multiple meshes and element types without concern for nodal or elemental interface matching.

3.11.2. Virtual Prototyping

Physical model building for evaluation can be time demanding and yield limited results. Using kinematic and dynamic studies on a design within a computer environment saves time, and the findings are frequently more helpful than experimental data from physical prototypes.

Physical prototyping frequently necessitates a significant amount of physical labor, not just to build the model's elements, but also to assemble them and attach the necessary instrumentation. Many of the same tests may be performed on a design model using kinematic and dynamic analytical approaches in virtual prototyping.

The intrinsic benefit of virtual prototyping is that it allows the engineer to fine-tune the design before creating a real prototype. When the prototype is finally built, the designer will almost certainly have more information with which to construct and test the model.

Physical models may give useful design data to engineers, but the time necessary to produce a physical prototype is lengthy and must be done frequently over process iterations. A second drawback is that the design is frequently updated over many iterations, therefore time is lost in the process when components are reassembled as a functional model.

In certain circumstances, the time spent on prototype development and testing yields less meaningful data than planned. One possible answer to the challenges of physical prototyping is virtual prototype of a design. Virtual prototyping uses computer-based testing to incorporate incremental design changes into the prototype model fast and efficiently.

Furthermore, using virtual prototyping, testing on the system or its components may be done in ways that would not be practical in a laboratory context. The equipment necessary to verify the operation of a small part in a system, for example, may disturb the system itself, depriving the engineer the exact information required to improve the design.

In addition, virtual prototyping may apply forces on the design that would be hard to apply in the laboratory. For example, if a satellite is to be built, the design should be subjected to zero gravity to accurately replicate its performance.

Engineers are increasingly doing kinematic and dynamic assessments on virtual prototypes because a well-designed simulation generates data that can be utilized to adjust design parameters and features that would not have been considered otherwise.

Kinematic and dynamic analysis methods apply physical rules to a computerized model to assess the motion of system components and evaluate the overall interaction and performance of the system as a whole. A mainframe computer was once required to execute the computations required to give a realistic motion simulation.

Microcomputers now have the processing speed and memory capacity to run such simulations on the desktop. The ability to purposely overload forces on the model is one advantage of kinematic/dynamic analysis software. The engineer may use the damaging testing data since the model can be recreated in an instant.

Every time the test was conducted, physical prototypes would have to be built and rebuilt. There are numerous instances in which physical prototypes must be built, but virtual prototyping analyzes may frequently make such scenarios more efficient and instructive.

3.11.3. Rapid Prototyping and 3D Printing

Rapid prototyping has been one of the most recent applications of CAD technology. Physical models have long been one of the greatest evaluation tools for influencing the design process. Regrettably, they have also been the most time-consuming and expensive step of the design process.

This issue is addressed by rapid prototyping, which combines CAD data with sintering, layering, or deposition procedures to generate a solid physical replica of the design or part.

The fast-prototyping industry is actively developing technologies to allow for the small-scale fabrication of genuine components, as well as molds and dyes, which may subsequently be utilized in subsequent traditional manufacturing procedures.

Because of these two objectives, the industry has grown specialized into two key segments. The first sector's goal is to develop compact fast prototyping equipment that will one day be as widespread in design offices as printers and plotters are now.

The fast-prototyping industry's second branch specializes in the fabrication of extremely accurate, structurally sound pieces for use in the manufacturing process. Because of the advancement of 3D printers, prototyping is now faster and cheaper than any prior rapid prototyping approach.

3.11.4. Additive Manufacturing

Computer-aided design (CAD) is used to develop the precise geometry necessary for the additive manufacturing (AM) technique of making items. Essentially, AM is the technique of making objects by successively adding layers of material.

In contrast, traditional machining involves the removal of material to create the final shape. AM enables the creation of components with complicated geometries without the use of specialized tools or fixtures, as well as the generation of waste material.

Combining CAD and AM technologies enhances the whole manufacturing value chain. Because of the geometric freedom provided by CAD/AM, traditional production constraints like as weight limitations, complicated curves, and intricate interior cavities may be significantly reduced.

This can be used for lightweight designs, lower part counts, or enhanced implant bone ingrowth. For example, the manufacturing path from CAD to completed product is characterized by high material usage and does not necessitate the storage of expensive castings, forgings, or molds, which is advantageous in mission-critical scenarios common in the military sector.

Aside from the potential for cost savings, AM's high material utilization is energy efficient and provides an ecologically benign manufacturing approach, confirming a competitive edge in sustainability. CAD is also an important component of 3D printing. This is the procedure for creating solid objects from CAD 3D model files.

It is connected to AM both technically and conceptually. There is equipment available that is branded as "desktop 3D printers," with prices ranging from $2000 and more. These printers replicate part designs made with 3D CAD software such as Solid works and Autodesk Inventor.

3.11.5. Collaborative Product Design

CPD is an interactive design and product development technique that involves designers from several geographical places working collaboratively to simultaneously design a product using potentially different versions of CAD software.

Through the use of open file format standards such as IGES, STEP, Virtual Reality Modeling Language (VRML), and Extensible Mark-up Language (XML), the development of specialist CAD software for collaborative product creation has now made this process much easier (XML).

CPD incorporates a number of engineering phases, including conceptual design, detailed design, engineering analysis, assembly design, process design, and performance assessment. CPD include gathering requirements, defining the overall aim and task, decomposing the overall task into hierarchical subtasks, assigning subtasks to engineering designers, solving the subtasks, synthesizing the sub solutions, and lastly giving the overall design solution.

The purpose of collaborating is to complete more work in a shorter amount of time by allowing a group to use the skill sets of all members of that group. CPD aids in the optimization of the product development process by eliminating mistakes between manufacturers and designers, enhancing product quality, lowering manufacturing costs, and offering a platform for collaborators from across the world to collaborate on a project concurrently.

CPD can assist in the early correction of faults committed during product design, lowering product development expenses, design iterations, lead times, and manufacturing costs.

Commercial systems such as engineering data management (EDM), PDM, product information management (PIM), technical document management (TDM), and technical information management (TIM) provide a structured way of efficiently storing, integrating, managing, and controlling data and engineering processes from design to manufacturing and distribution.

CPD software can be utilized as a cloud-based service or as a SaaS. (software-as-a-service). The CAD software package is installed at each individual workstation in the cloud-based alternative, and each individual designer shares his or her work with others through the cloud (the cloud is a global network of computers communicating with each other over the Internet).

The advantage of this technique is that communication speed is increased since engineers only communicate critical data over the cloud. Each workstation must have CAD software installed. The program is not installed on each individual workstation in the SaaS option, but is instead subscribed to as a service (typically by paying a subscription fee) from a business that maintains the software on the cloud.

Using the SaaS approach may provide several challenges, such as explicitly declaring ownership of any data or designs, ensuring software update continuity, and contingency planning for vendor changes. The most significant advantage of SaaS is that just one copy of the program is required, thus lowering maintenance expenses.

To minimize costs even further, certain CAD software may integrate remote desktop applications. A remote desktop application allows programs or apps to be executed remotely while still being accessible locally.

3.11.6. Product Life-Cycle Management

CAD is an essential component of PLM. PLM is a holistic approach to innovation, new product development, launch, and product life cycle management (PDM) from concept to end of life. When combined with CAM, whole product life cycles may be investigated.

Typically, product functionality, affordability, and sustainability are evaluated. PLM may be viewed as a store for all product information as well as a communication method among product stakeholders such as marketing, engineering, manufacturing, and field service. A typical PLM system allows product information from marketing and design to be combined and then transformed into production information.

The PLM methodology was originally used in mission-critical applications such as aerospace, medical, military, and nuclear. Configuration management and electronic data management systems spawned PLM. PLM is currently used by makers of industrial machinery, biomedical equipment, consumer electronics, consumer goods, and other engineered items.

PLM focuses on technical areas of the company such as engineering, design, and manufacturing, as well as integration with enterprise resource planning (ERP) system planning and order information.

3.11.7. Product Data Management

The software system of standard CAD systems has continued to grow toward integration with a PDM system. A PDM system typically comprises of centralized software that controls access to and manages product data and process-related information. This data comes from CAD data, models, part information, production instructions, requirements, notes, and papers.

The ideal PDM solution serves business-specific needs and is accessible by many applications and teams within an organization. As global organizations expand, the CAD techniques and technology discussed in this chapter will play an increasingly important role in the evolution of these interconnected systems.

3.12. PRODUCT DESIGN FOR MANUFACTURING AND ASSEMBLY

Product design approaches are changing dramatically at all stages of the new product development process. Over the next decade, these innovations will have a substantial influence on how all goods are developed as well as the development of associated production processes.

The rapid pace of technological development has produced a dynamic scenario that most corporations have found challenging to manage. Some analysts openly state that if there is no new technology over the next five years, corporate America may begin to catch up.

Up-front technology, engineering, and design techniques that promote and support a wide range of new product development methods are critical to attaining benchmark speed to market, cost, and quality. These procedures must incorporate contemporary manufacturing technology as well as component parts intended for simplicity of assembly and parts that can be produced utilizing low-cost manufacturing techniques.

When the designs of machines and the manufacturing methods that generate those machines are consistent, optimal new product design happens. The obvious purpose of every new product development process is to generate revenue by transforming raw materials into finished goods. This may appear to be a simple task, but it must be done efficiently and affordably.

Many businesses may not know how much it costs to create a new product until it is in production. Rule 1: At the outset of the project, the product development team must be given a cost objective. This cost will be referred to as the unit manufacturing cost (UMC) objective.

Rule 2: This goal cost must be held accountable by the product development team. DFM&A tools assist the development team in minimizing the number of individual components that comprise the product while also ensuring that any new or residual pieces are easy to handle and insert throughout the assembly process.

DFM&A promotes the integration of components and processes, which helps to minimize assembly labor and costs. DFM&A activities include projects to reduce overall product development cycle, manufacturing cycle, and product life-cycle costs. Furthermore, DFM&A design programs encourage teamwork as well as supplier strategy and commercial concerns at an early stage in the product development process.

The DFM&A process is divided into two parts: design for assembly (DFA) and design for manufacturing (DFM) (DFM). DFA denotes the labor component of the product cost. This is the time required to turn the new design into a customer-ready product. DFM is the new product's material and tooling side.

DFM breaks down the parts manufacturing process into its most basic components, such as the type of equipment used to create the part and fabrication cycle time, and determines a cost for each functional step in the process. Before beginning the new product design process, the program team should utilize DFM tools to determine the material target cost.

Manufacturing expenditures are borne during the project's early design phase. Many studies have revealed that up to 80% of the cost of a new product is fixed in stone during the product's initial drawing release phase. Many firms struggle to make adjustments to their new product development process.

3.12.1. What Is DFM&A?

DFM&A is not a panacea. It is a tool that, when utilized correctly, may have a significant impact on the design philosophy of any product. DFM&A's major purpose is to reduce product costs by reviewing product design and structure during the early idea stages of a new product.

DFM&A also contributes to improvements in serviceability, dependability, and product quality. It reduces overall product costs by focusing on assembly time, part cost, and the assembly process early in the product development cycle. A product's life cycle begins with the definition of a collection of product demands, which are subsequently converted into a set of product conceptions.

Design engineering refines these product concepts into a precise product design. Given that the product will most likely be in production for several years beyond this point, it seems reasonable to take time during the design process to question, "How should this design be put together?" This will make the remainder of the product's life much easier after the design is finished and passed off to manufacturing and servicing.

To be fully effective, the DFM&A process should begin within the project's idea development phase. True, applying DFM&A during the frenetic design process will take time, but the advantages readily worth the extra work.

The development team use DFM&A as a tool to drive particular assembly advantages and highlight downsides of various design options, as assessed by attributes such as total part count, handling, and insertion difficulties, and assembly time. DFM&A turns time into money, which should be the standard metric for comparing various designs or redesigns of an existing concept.

The early DFM&A study gives the product development team a baseline against which to make comparisons. By keeping an itemized count of each part's influence on the entire assembly, this early study will assist the designer in understanding the precise pieces or concepts in the product that require further refinement.

Once a user has mastered a DFM&A tool and the concepts have become second nature, the tool remains an effective way of consolidating what is now second nature to DFA veterans and assisting them in presenting their ideas to the rest of the team in a common language: cost.

Because of the rigorous design limitations imposed on the auto valve project, engineering believes they will lack the resources to implement the design for manufacture and assembly. Furthermore, this software is heavily time-bound.

The project's budget has already been authorized, as have other components of the program that must be completed on schedule in order to meet top management's financial objectives.

3.12.2. DFM Analysis

The DFM analysis was used to calculate the cost of the fabricated item. The foundation, for example, is machined from solid aluminum bar stock. The foundation has 11 separate holes drilled into it as specified, with 8 of them requiring tape. According to the DFM study (see Table 3), it takes 17.84 minutes to manufacture this item from solid bar stock.

In batches of 1000 components, the final machined base costs $10.89. The optimal method for doing a DFM analysis is as follows. In the instance of the foundation, the solid geometry was generated in Matra Data's Euliked computer-aided design (CAD) system by the design engineer.

The design engineer then provided the solid database as an STL file to the manufacturing engineer, who opened the STL file in Solid View. The mechanical engineer was able to get all of the dimensioning and geometry inputs required to complete the Boothroyd Dewhurst DFM machining analysis of the basic part using Solid View.

Solid View also enabled the mechanical engineer to take cut portions of the part and step through them to check that no production regulations were broken. The STL file output format is now supported by all major CAD suppliers. Many new CAD viewing tools, such as SolidView, are now available for $500–$1000. These applications handle STL or IGS files.

The objective is to connect all of the early product development data so that each member may provide quick, accurate feedback on the design at its earliest stage. In this case, the mechanical engineer spent 20 minutes pulling the STL files into Solid View and doing the DFM analysis.

Engineering has previously complained that DFM&A takes too much time and slows down the design team. Following the producibility rule, the mechanical engineer examines the base as a die casting part. Many of the part characteristics may be molded into the part by creating the base as a die casting.

This net form die cast design eliminates most of the machining that was previously necessary. The die-cast portion will still need to be machined. According to the DFM die casting study, the base casting would cost $1.41 and the mold would cost $9050. Table 4 compares the two ways of manufacturing.

3.12.3. DFM&A Road Map

The team leader should foster and cultivate a creative environment. It is critical to construct a product development team that has the expertise to make the correct decisions, the capacity to carry them out, and the perseverance and devotion to see the project through to completion.

Although these abilities are priceless, it is equally important that these professionals be given as much latitude as possible to sprout innovative solutions to design problems as early in the product design cycle as feasible. The product development team is in charge of both the product design and the development process.

The DFM&A approach is most successful when applied by a multifunctional team, with each member bringing his or her own area of expertise to the product design process. For DFM&A to be most effective, the team should embrace group dynamics and the collective decision-making process.

Clearly, there might be negatives to multidisciplinary teams, such as managing too many viewpoints, trouble making choices, and other variables that may extend the product development cycle in general. However, after a team has worked together and understands individual duties, there is a lot to be gained by using the team method.

Working groups can pool their particular talents, abilities, and expertise to bring greater resources to bear on an issue. Group discussions help people grasp issues, ideas, and potential solutions from a range of perspectives.

Because people are more driven to support and carry out a choice that they helped develop, group decision making leads to greater commitment to decisions. Individuals can enhance their present abilities and acquire new ones in groups.

3.12.4. Why Is DFM&A Important?

DFM&A is a valuable tool in the design team's arsenal. When utilized correctly, it may provide fantastic effects, not the least of which is that the product will be simple to build!

The most favorable result of DFM&A is a reduction in part count in the assembly, which simplifies the assembly process, lowers manufacturing overhead, reduces assembly time, and increases quality by reducing the chances of introducing a defect. Labor content is also minimized since fewer parts are used, the assembling activities are fewer and easier. Another advantage of lowering part count is that the product development cycle is shorter since there are fewer parts to design.

The ideology promotes design simplification and the use of conventional, off-the-shelf elements whenever possible. When adopting DFM&A, a fresh emphasis is placed on designing each item so that it may be manufactured inexpensively by the chosen manufacturing method.

3.13. CONCLUSION

In the conclusion of this chapter, it discussed about basics of mechanical engineering. This chapter also, discussed about the basic concepts such as force, work, energy, power, and friction. It also provides highlights on the significance of computer – aided design, and its software. Towards the end of the chapter, it discussed about the 2D and 3D graphics, various applications of CAD, CAD standards and translators, and product design for manufacturing.

REFERENCES

1. (2009). Fundamentals of Mechanical Engineering. [e-Book] IDC Technologies. Available at: https://www.idc-online.com/downloads/ME_r5.pdf (accessed on 1 July 2022).
2. Abrams, L., Altschuld, J., Lilly, B., & Mendelsohn, D., (n.d). Introduction to mechanical engineering: A course in progress. 2012 ASEE Annual Conference & Exposition Proceedings. [online] Available at: https://peer.asee.org/introduction-to-mechanical-engineering-a-course-in-progress (accessed on 1 July 2022).
3. Clifford, M., (2014). An Introduction to Mechanical Engineering: Part 2. [online] Available at: https://www.taylorfrancis.com/books/mono/10.1201/b13328/introduction-mechanical-engineering-part-2-michael-clifford (accessed on 1 July 2022).
4. Dominico, S., (2020). Introduction to mechanical engineering. J. Paulo Davim. Chemie Ingenieur Technik, 92(6), 788–800. [online] Available at: https://onlinelibrary.wiley.com/doi/10.1002/cite.202070607 (accessed on 1 July 2022).
5. Kutz, M., (2015). Mechanical Engineers' Handbook, Volume 2: Design, Instrumentation, and Controls. PDF Drive. [online] Pdfdrive.com. Available at: https://www.pdfdrive.com/mechanical-engineers-handbook-volume-2-design-instrumentation-and-controls-d178659083.html (accessed on 1 July 2022).
6. Seeler, K., (2014). System Dynamics. [online] Available at: https://link.springer.com/book/10.1007/978-1-4614-9152-1 (accessed on 1 July 2022).

CHAPTER 4

ENGINEERING COMMUNICATION & ETHICS

CONTENTS

4.1. INTRODUCTION

Engineering is the practice of employing technology to create an efficient mechanism that speeds up and simplifies operations while using limited resources. Ethics are socially accepted principles that correspond to human moral standards. An ethical engineer can make a positive difference in society.

The study of engineering ethics, and the application of such ethics in engineering by engineers, is essential for the betterment of society. Engineering ethics is the study of morally desirable decisions, policies, and ideals in engineering practice and research.

Communication Engineering is a branch of electrical and computer engineering that focuses on communication. This field is concerned with improving telecommunication systems. Communication engineers are in charge of telecommunication equipment design, installation, and maintenance. It's a broad field that overlaps with Civil, Electronic, and System Engineering.

Communication engineers are also responsible for high-speed data transfer. To design the Telecom Network Infrastructure, these specialists use a variety of transport media and equipment. They're also useful for wireless communication.

This field was established in the late 1800s. It was, however, not very popular at the time. However, this discipline has grown in popularity as a result of the increased usage of communication platforms. This article contains comprehensive information on Communication Engineering study resources and lecture notes.

4.2. ETHICS

"Ethics" comes from the Greek word "ethos," which means "character." Ethics are a collection of norms or principles that are widely seen as standards of good and bad, or right and wrong, and are typically enforced by an external entity, such as a community or a profession.

Ethics is important in our daily life because we will face an ethical question or problem at some point in our professional or personal life, such as what is our level of responsibility for safeguarding another person from harm, or whether or not one should tell the truth in a certain situation.

Ethics can be defined as a society's proposed or acknowledged norms of conduct for a specific class of human actions, or a specific group or culture.

Ethics are defined differently by different people. They may or may not change depending on the situation.

Ethics and morality are closely linked terms. Where earlier it would have been more accurate to speak of moral judgments or moral principles, it is now usual to refer to ethical judgments or ethical principles. These applications go beyond the definition of ethics.

Previously, the phrase referred to the field of study, or branch of investigation, that had morality as its subject matter, rather than to morality itself. Ethics is similar to moral philosophy in this regard. A person who carefully follows a set of ethical rules may lack moral integrity, whereas a person who occasionally violates ethical norms may have high moral integrity (Figure 4.1).

Figure 4.1. Ethics plays an important role in engineering.

Source: Image by the blue diamond gallery.

Duty ethics, right ethics, virtue ethics, and other ethical philosophies are examples. Utilitarianism is an excellent example of ethics. Utilitarianism is a philosophy that states that the greatest good is the happiness or pleasure of the largest number of people in a community.

According to this theory, an action is ethically correct if its results lead to people's happiness, and it is morally wrong if the activity causes them to be unhappy. This philosophy looks beyond one's own interests and considers the interests of others.

4.2.1. Ethics in Engineering

Ethics are standards that are followed based on a person's sense of moral obligation. Engineering ethics is the study of connected problems about moral values, character, policies, and relationships of persons and organizations involved in technical activity.

An engineer, whether working alone or for a firm, must deal with ethical difficulties, which arise in situations such as product conceptualization, design, and testing issues, or manufacturing, sales, and service issues. During supervision and group work, moral questions arise as well.

Because an engineer's ethical decisions and moral values have an impact on the products and services, it is important to discuss them – how safe they are to use, the company and its shareholders who believe in the company's goodwill, the public and society who trust the company regarding the benefits of the people, the law, which cares about how legislation affects the profession and industry, the job and his moral responsibilities and about the law, the job and his moral responsibilities and about the law.

4.2.2. Ethics and the Engineer

Why is ethics such an important component of professional life, particularly in the field of engineering? Consider what a professional is to understand the significance of ethics in the professions. Even in established professions like medicine, law, accounting, and engineering, the term 'professional' is difficult to define. However, there is general agreement on common features that all occupations share. As a result, a professional:

- possesses specialized skills and knowledge;
- has acquired such information and abilities over the course of a long period of training and study, and continues to maintain and update them throughout his or her career;
- possesses significant power over individual clients and society as a result of its specialized expertise;
- is a member of a professional organization that oversees their work; and
- and, as part of such self-regulation, follows ethical norms overseen by the professional body.

Professionals' expertise and the domains in which they execute it provide them the power to enhance people's lives or to wreak considerable harm. This is undoubtedly most clear in the case of doctors, whose actions have the potential to save or end lives, as well as effect quality of life in a variety of other ways (Figure 4.2).

Figure 4.2. Code of ethics.

Source: Image by the blue diamond gallery.

A patient needs to know that a medical professional will treat them ethically informed judgment, acting solely with their agreement, keeping confidentiality, and seeking their best interests, among other things. While medical professionals' activities usually have a direct impact on individual patients, engineering professionals' decisions have the ability to affect the well-being of hundreds or thousands of individuals.

Because of the power that their skills confer, society invests a high level of faith in professionals to use those skills properly. A commitment to apply expertise for the public benefit is thus shared by all professions. Because the professional's adherence to ethical principles is a vital aspect of the exercise of excellent professional judgment, ethics plays a critical role.

By doing so, the professional wins the public's trust and demonstrates why that trust should be maintained. In short, being a professional grants you considerable powers over others, whether it's access to information about them or the ability to influence their needs and desires. Those privileges come with significant obligations; thus, professions, and professional bodies must constantly earn the right to be entrusted with them by demonstrating that they are carried out ethically.

Engineers operate in a variety of fields, but they all have the potential to have a significant impact on social well-being. Engineers play large roles in two extremely important facets of human life, as Richard Bowen has pointed out. Engineers, on the one hand, can propose solutions for better water resource management and treatment.

Such engineers have the potential to do immense good in a world where a major section of the worldwide population (1.1 billion according to the World Health Organization in 2004) lacks access to safe drinking water. Engineers, on the other hand, play an important role in the defense industry. In serving to defend people from aggressors, this action has the potential to accomplish a lot of good, but guns can also be used to do a lot of harm.

The honor of having the abilities and knowledge to make such a significant contribution to such vital areas of life certainly necessitates exercising sound ethical judgment when doing so. Engineers, on the other hand, have a variety of direct and indirect effects on individual and societal well-being. When a person walks across a bridge, they need to know that engineers have properly balanced the vital imperative of safety with the demands for cost-effective construction and a pleasing esthetic outcome.

A mining project's site necessitates sound judgment, taking into account environmental and other factors as well as meeting technical and commercial criteria. Engineers must consider the sustainability of their processes since they employ material and energy resources in the manufacture, packaging, and distribution of items they design and make. All of these ramifications must be understood by responsible engineers, who must act correctly in light of them.

Clearly, engineering specialists must be trusted in a wide spectrum of human endeavors. Engineers, like any other profession, require sound ethical judgment. How does ethics, on the other hand, differ from common sense?

There are numerous examples that demonstrate that competent people with sound judgment can disagree on ethical issues. Modern electronic gadgets that allow surveillance are frequently touted as beneficial in combating terrorism, but opinions differ on whether the resulting invasion of privacy is justified.

Some people believe that producing wind power is an environmentally beneficial solution to meet electrical needs, while others believe that the enormous turbines' influence on the landscape is harmful to the ecosystem. The case studies in this guide demonstrate the limitations of common sense when it comes to dealing with engineering ethics (Figure 4.3).

Figure 4.3. Ethics in different field.

Source: Image by the blue diamond gallery.

Do such challenging examples demonstrate that ethical questions are purely subjective, with no right or wrong answers? They simply demonstrate that the correct answer is not always evident, as even the most difficult issues have 'wrong' solutions. Keeping everyone in their houses 24 hours a day is not an acceptable approach of ensuring security. Mineral extraction that is completely unregulated should not be permitted. The goal of this guide and the Statement of Ethical Principles is to demonstrate that it is possible to identify significant considerations for making ethical judgments and to utilize reason in deploying those considerations.

4.3. WHY STUDY ENGINEERING ETHICS?

Engineering ethics should be researched because it is crucial in ensuring the safety and usefulness of technological goods as well as giving meaning to engineers' efforts. It's also complicated in ways that necessitate thoughtful consideration throughout a career, beginning with the pursuit of a degree. What specific goals should govern the study of engineering ethics, beyond these broad observations?

The direct aim, in our opinion, should be to improve our ability to deal successfully with moral complexity in engineering. As a result, studying engineering ethics improves our ability to reason effectively and thoughtfully about moral issues. To use ethical terminology, the overarching purpose is to increase moral autonomy.

Autonomy implies self-determination, yet not all independent ethical thinking qualifies as moral autonomy. Moral autonomy can be defined as the ability and habit of rationally thinking about ethical concerns based on moral concern and commitment. This foundation of moral Professionalism values responsiveness stems mostly from the teaching we acquire as children in being attentive to the needs and rights of others as well as our own.

When such teaching is lacking, as it frequently is with severely mistreated children, the tragic result might be an adult sociopath with no sense of moral good or wrong. Regardless of how independent their logical thinking about ethics may be, sociopaths (or psychopaths) are not morally autonomous.

4.4. ENGINEERING COMMUNICATION

The use of communication devices dates back to the late 1800s. It has gained popularity in recent years as a result of changes in communication. Telecommunications has grown in importance around the world. TV, radio, the Internet, satellite, and other forms of communication are all used in communication engineering. Information security, telephone systems, and computer networks are all areas where Communication Engineers become experts.

Any living being that coexists with another must exchange some information. When the need for information exchange occurs, some kind of communication should be available. The necessity for communication is unavoidable, regardless of the means of communication, which can include gestures, signs, symbols, or a language. Human communication relies mostly on language and gestures, whereas animal communication relies heavily on noises and actions. When a message must be transmitted, however, communication must be established.

4.4.1. The Process of Communication

Figure 4.1 depicts the basic model of the communication process. The communication starts with a problem perspective. The sender wishes to convey information to the receiver. When information is transformed into a systematic collection of symbols (language) that expresses what the sender desires to relay, this is called encoding.

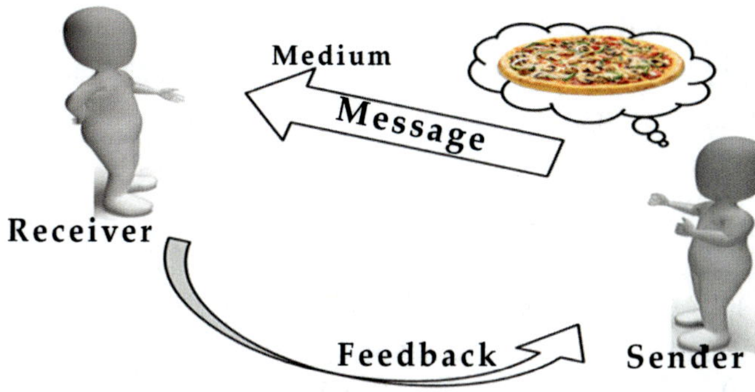

Figure 4.4. Process of communication.

Source: Image by commons.wikimedia.org.

A message is the result of encoding. The message's format is determined by the communication channel. The channel is the path that the communication will take from sender to receiver. The message could be a written report, a face-to-face conversation, a phone call, or a computer network broadcast.

When a message reaches its intended recipient, it is interpreted in light of the recipient's prior experience. The receiver's feedback to the sender indicates how the message was delivered. Feedback allows the sender to determine whether the message was received and whether it had the desired effect – the receiver's adequate understanding of the situation and transfer of meaning.

The communication process faces a number of challenges due to psychological and semantic interferences (barriers). The following psychological interferences are related with the communications environment:

- Failure to specify the purpose;
- Lack of regard for one another (sender and receiver);
- Either party's preconceptions;
- Either party's preoccupation;
- Failure to identify the most effective channel;
- The sender's identity is unclear;

- The sender/receiver is closed to criticism; and
- Emotions aren't taken into account.

Professional success in engineering continues to be dependent on excellent communication skills. Communication competency, on the other hand, has shifted to encompass communicating with technology, the effects of the global market and social milieu on communication, and mutual appreciation and respect for disciplinary and cultural diversity.

These minor developments necessitate a rethinking of communication teaching in order to equip engineers to communicate effectively in the global workplace with a broad, multinational audience. This text aims to demonstrate how communication across disciplines (CID) may help engineers prepare for global involvement and citizenship. It employs metaphors to demonstrate how current CID work stresses communication as a tool for achieving professional objectives. By positioning communication competency as a forceful, consequential interaction, it proposes the metaphor of voice for (re)imagining a larger approach to CID that will prepare students for communication in the global workplace.

4.5. SCOPE OF ENGINEERING COMMUNICATION AND ETHICS

Engineering is defined as "the creative application of scientific principles to design or develop structures, machines, apparatus, processes, or works utilizing them singly or in combination; or to construct or operate the same with full cognizance of their design; or to forecast their behavior under specific operating conditions."

As a result, engineering is concerned with turning research into practical goods for human comfort. Engineers conduct engineering, and what they accomplish has far-reaching consequences for others. The ability and obligation of an engineer to appraise his judgments in the perspective of society's overall well-being is referred to as engineering ethics. It is the study of moral difficulties that engineers and engineering organizations face while making important decisions.

Engineering research and practice demand that the work at hand analyzes all of the advantages and disadvantages of a particular action and its implementation. On the basis of their members' extensive experience, professional engineering groups such as IEEE, ASME, and IEI have developed comprehensive ethical codes pertinent to their respective professions. Independent groups such as the National Society of Professional Engineers (NSPE) have developed value-based ethical codes that apply to all engineering professions.

Engineering ethics is mostly taught at academic institutions through a variety of case studies aimed at raising awareness among engineering students of all disciplines. Students improve knowledge and assessment skills of the anticipated consequences of their future decisions on moral and ethical grounds by studying engineering ethics. Several variables influence engineering ethical standards:

- Engineering as a human endeavor with far-reaching consequences is a noteworthy factor;
- Ethical difficulties make engineering judgments comparatively tough to make;
- A key concern of an engineer is the risk and safety of people as a social obligation;
- Technological advancement can be very demanding on engineering talent in a worldwide environment; and
- Moral principles and responsible conduct will play a crucial role in decision making.

Engineering ethics is taught as part of an engineering degree to help students prepare for their future careers. Engineers who study ethics benefit in particular because they gain clarity in their understanding and thought about ethical concerns and the practice in which they arise. The study of ethics aids students in developing transferable communication, reasoning, and reflection abilities.

These qualities enable students to engage in other elements of the engineering program, such as group projects and job placements. When we talk about the scope of communications engineering, we're talking about the application of science and math to real-world communications difficulties. Can one envisage a world without a mobile phone, a laptop (computer), a television (TV), tablets, a digital watch, internet banking, ATM cards, Wi-Fi, internet access, a microwave, and a variety of other gadgets and communication systems? It is a nearly impossible task.

Electronics and Communication Engineering have made all of this possible (ECE). As humans look to the future, robots will become an increasingly significant part of our lives, and embedded electronics, a subfield of electronics, will play a key role in this. Because the world is progressing in the realm of technology and advancements, the scope is very broad.

- It has applications in practically every industry, including oil, energy, and agriculture, because electronics and computers are used in so many other critical sectors of the economy.
- Electronic engineers are employed by defense, space, and other large-scale research companies in the development and design of communications and signal processing systems and devices.
- ECE's use in numerous specializations such as VLSI Design, Embedded Systems, Communication Engineering, Signals, and Systems, Microwave Communications, and so on.

CE is one of the largest and fastest developing branches of engineering, and it plays a critical part in the Technology Revolution. The current global technological revolution is transforming the world and providing challenging challenges for engineers in particular. CE is a branch of EE that is studied at the graduate level.

The Primary Objective of CE is to: The main goal is to create smaller, smarter, and multi-functional goods. Digital information processing is enabled by electronic devices' ability to operate as switches. It includes a variety of applications that make our lives easier and more fun, such as TV, radio, computers, and telephony. No one could have predicted that when Alexander Graham Bell built the telephone and Marconi produced the radio in the nineteenth century, these two separate technologies would combine to create another fantastic product, the cell phone, which is now a part of everyone's life. Ted Hoff invented the commercial microprocessor at Intel in 1969, paving the way for the personal computer's growth.

4.6. PROFESSIONAL CODES OF ETHICS

A code of ethics outlines how professionals should pursue their common goal in order to accomplish the best for themselves and people they care about. The purpose of the code is to insulate each professional from certain pressures (e.g., the pressure to make corners to save money) by ensuring that most other members of the profession will not take advantage of them.

A code is a method of resolving a coordination issue. A professional has responsibilities to his or her employer, customers, and other professionals—colleagues—with specific reciprocity expectations. Codes of ethics are adopted by engineering societies to express the rights, duties, and obligations of members of the profession. Codes of ethics are not only found in professional organizations, but also in universities.

A code of ethics provides a foundation for professional ethical judgment. No code can be completely comprehensive and cover every ethical circumstance that a professional engineer (PE) might face. Rather, norms are used to help people make ethical decisions. No new ethical principles are established by ethical codes.

They merely restate ideas and standards that have long been acknowledged as best practices in engineering. A code presents these principles in a way that is consistent, comprehensive, and easy to understand, allowing the engineer to apply them to the unique scenarios that arise in professional practice. Finally, a code establishes professional roles and responsibilities (Figure 4.5).

Figure 4.5. Professional codes of ethics.

Source: Image by Pxhere.

Employees who are being coerced by their employers to do anything unethical, or who are accusing their employers or the government of unethical behavior, should have their rights protected by professional societies. Employees can utilize the codes of professional associations as a defense against their employers, therefore they are useful.

If professional societies' rules of ethics are to be meaningful, their support for engineers' defense is required when ethical infractions are identified. However, because not all engineers belong to professional societies, and engineering societies are generally weak, the pressure that these groups may exert is limited.

4.7. THE PROFESSIONAL APPROACH TO ENGINEERING ETHICS AND CODES OF CONDUCT

The premise that engineering, like medicine and accounting, is a profession lies at the heart of what is known as the professional approach to engineering ethics. A profession can be defined as an occupation with a variety of distinct characteristics.

The utilization of specific knowledge and abilities, as well as a (legal) monopoly on the performance of the occupation, are aspects that are frequently cited in this regard. For example, the latter would imply that not everyone may call themselves engineers or perform engineering job. Exercising a profession is frequently associated with the legal protection of titles and university or college degrees.

The idea of a profession is also linked to the idea that evaluating professional work and determining whether or not a professional has done his or her job (in)competently or (im)properly can only be done by peers, because they are the only ones with the knowledge and skills to apply the appropriate standards of judgment.

While engineering undoubtedly exhibits some of the above characteristics, it does so to a lower extent in most countries than, say, medicine. Engineers in the United States, for example, must be licensed as engineers to undertake engineering work, which entails certain checks on their knowledge and skills, although there is an exemption for engineers who work in industry, which is likely the majority of engineers.

There are no licensing or registration requirements in many other nations, or only for a small number of engineers. While most doctors are subject to disciplinary law and must appear in front of a disciplinary court in cases of patient complaints, this is not always the case in engineering (with some exceptions). The majority of engineering job is exempt from disciplinary laws.

A profession is sometimes thought to be committed to a concept of serving society and particular moral purposes, in addition to these more descriptive characteristics. According to conceptualist approaches to professional ethics, each profession is associated with a certain goal that is determined internally by that profession (Davis 1998).

The eventual result in medicine would be something like human health. In engineering, one might consider human well-being to be a goal to be achieved. However, one distinction between medicine and engineering appears to be that, whereas doctors have specialized information that aids in the definition of health, engineers lack such knowledge.

Furthermore, while medicine appears to be the primary profession concerned with human health, engineering does not appear to hold the same place in terms of human well-being. Michael Davis, one of the primary authors in the professional tradition in engineering ethics, has offered a more historical approach to defining a profession as an alternative to conceptualist methods. "a number of individuals in the same occupation voluntarily organized to earn a living by openly serving a certain moral ideal in a morally-permissible way beyond what law, market, and morality would otherwise require" he defines a profession as (Davis 1998: 417).

The fundamental criterion for determining whether an occupation qualifies as a profession, according to this definition, is whether there is a voluntary commitment to a moral ideal. As a result, for Davis, the aforementioned descriptive characteristics are irrelevant when it comes to engineering as a vocation.

Davis believes that engineering is a profession in most countries because engineers have committed to particular moral values, such as engineering codes of ethics. Engineers and engineering associations have developed codes of ethics in numerous nations. The oldest is most likely the Smeatonian Society's code of ethics, which was established in 1771. (Van de Poel and Royakkers 2011).

American engineering societies, such as those for civil engineering and mechanical engineering, created codes of ethics in the early twentieth century, as part of their quest to be recognized as real professionals. While older codes frequently emphasized politeness and correct behavior toward other professions, clients, and employers, engineering codes of ethics evolved to emphasize responsibility to the public and society, particularly after WWII.

Many codes of ethics now specify that "engineers should priorities the public's safety, health, and welfare" (NSPE 2007), implying that this moral commitment takes precedence over other moral obligations that engineers may have, such as to clients or employers. According to Davis, this appears to be the kind of moral ideal that defines a vocation.

Most existing codes of ethics address three basic types of obligations and responsibilities for engineers: 1) competent and integral performance of the profession, including upholding moral values such as honesty, integrity, competence, independence, and impartiality; 2) acting as loyal and trustworthy agents to their clients and employers, including values such as loyalty, confidentiality, and faithfulness; 3) performing specific public obligations, such as prioritizing public safety, health, and welfare, serving the public interest, sustainability, and social responsibility.

Engineering societies' codes of ethics are usually aspirational or advisory in character. They express the ideals that engineers hold dear, and they frequently attempt to offer ethical counsel to practicing engineers. They are not disciplinary in most circumstances, which means they do not have a (semi)legal status.

Engineering codes of ethics are also not actively enforced in most nations, in the sense that people can be prosecuted for breaking the code and, in certain cases, expelled from the profession. Nonetheless, some engineers have been expelled from engineering societies for (supposedly) breaching the code of ethics, particularly in Anglo-Saxon countries.

4.8. ETHICAL THEORIES

Ethical philosophy emerged throughout history, with Greek, Hindu, Chinese, and Islamic philosophers developing ethical ideas independently.

Personal ethics are anchored in religious beliefs for many people, although this is not the case for everyone. There are many ethical people who are not religious, and there are innumerable examples of religious people who are not ethical. While the ethical principles we'll cover are filtered through a religious tradition, they've become cultural norms in the modern world, and they're widely accepted regardless of where they came from.

Engineers apply ethical ideas to ethical challenges they face in order to analyze and solve them. To build practical ethical problem-solving approaches, we must first examine numerous ethical theories in order to establish a decision-making framework.

When solving a problem in most engineering classes, there is usually only one theory to consider. There are numerous theories that will be addressed when studying engineering ethics. The fact that there are so many theories does not suggest a lack of theoretical grasp of ethics.

Rather, it reflects the complexity of ethical dilemmas and the variety of ethical problem-solving approaches that have evolved over time. Because each theory emphasizes various parts of an issue, having many theories to apply enriches the problem-solving process, allowing problems to be looked at from diverse perspectives. The basic ethical issue-solving technique analyzes the problem using several ideas and approaches before attempting to discover the optimal answer (Figure 4.6).

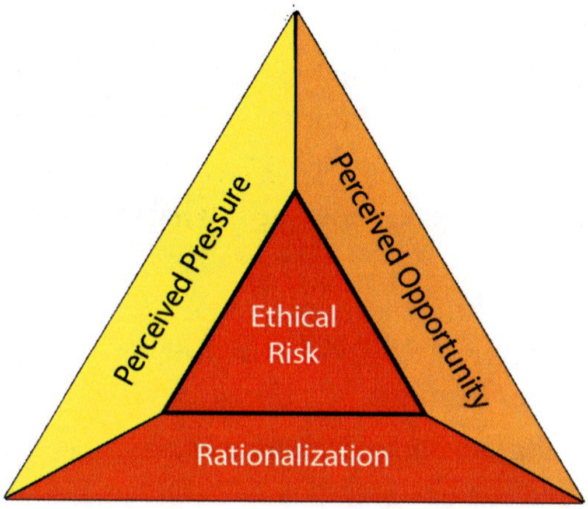

Figure 4.6. Ethical risk.

Sources: Images by commons.wikimedia.org.

An ethical theory defines words consistently and connects ideas and situations in predictable ways.

There will be four ethical theories discussed here, each with its own interpretation of the most essential moral concept:

- Utilitarianism aims to maximize utility, which is described as a balance between the positive and negative consequences of an action, taking into account the effects on everyone involved.
- Duty ethics asserts that certain responsibilities must be fulfilled regardless of whether they result in the greatest good. Rights ethics emphasizes that we all have moral rights, and any action that violates these rights is ethically unacceptable.
- Virtue ethics views behavior that reflects positive character traits (virtues) as proper and actions that display undesirable character traits as wrong (vices)

If valid conclusions are to be reached from any comprehensive investigation of an ethical situation, numerous ideas must be considered.

4.8.1. Utilitarianism

Utilitarianism advocates for behaviors that promote human well-being. The emphasis in utilitarianism is on maximizing the well-being of society as a whole, rather than on maximizing the well-being of individuals, making it a collectivist approach.

Many types of engineering analysis, such as risk–benefit analysis and cost–benefit analysis, rely on utilitarianism. However, as appealing as the utilitarian principle may appear, it has major flaws:

- What is best for everyone may be harmful to a single person or a group of people
- Its implementation is highly dependent on knowing what will result in the greatest good. It is frequently impossible to predict the exact implications of a decision.
- Maximizing societal benefit necessitates guesswork and the chance that the best guess is incorrect.

Two Basic Types of Utilitarianism:

Individual actions should be judged depending on whether the best was produced in a given situation, and regulations should be ignored if doing so will lead to the best, according to act utilitarianism. According to rule utilitarianism, moral principles are the most important, and while following them may not always maximize good in a given scenario, following them will ultimately lead to the greatest good.

Dam construction is an example of how this principle has been applied in the last century: Dams frequently serve society by providing reliable drinking water supplies, flood control, and recreational opportunities. These benefits, however, frequently come at the expense of individuals who reside in areas that will be flooded by the dam and must relocate or lose their land. Utilitarianism aims to strike a balance between societal and individual demands, with a focus on what will benefit the greatest number of people.

Cost-benefit analysis is a common approach in engineering analysis, particularly when determining if a project is feasible. Only projects with the greatest benefit-to-cost ratio will be implemented. The utilitarian goal of maximizing the total good is the same as this idea.

While the expenses of most projects are usually easy to forecast, the benefits obtained from them are typically more difficult to predict and assess. From a cost–benefit standpoint, it may appear that building a dam is a fantastic idea. Other problems, such as whether the advantages balance the loss of a living space or the loss of an endangered species with no current economic worth, will not be considered in this study.

Finally, it is critical to ensure that those who profit will also be responsible for the expenditures.

4.8.2. Duty Ethics and Rights Ethics

Duty ethics and rights ethics are closely related. These views claim that activities that respect an individual's rights are good. The welfare of society as a whole is not the primary moral factor here.

Moral activities are those that can be written down on a list of responsibilities: be honest, don't cause others to suffer, be fair to others, and so on.

These behaviors are our responsibilities because they demonstrate respect for individuals, autonomous morality, and universal ideals. The ethically proper moral activities are clear once one's responsibilities are realized. In this definition, ethical acts are the result of one's duty fulfilment. People's fundamental rights, such as their right to life, liberty, and property, are upheld by rights ethics, which others must respect.

Both obligation ethics and rights ethics have the same goal: individual persons must be respected, and behaviors that maintain that respect are ethical. People have responsibilities under responsibility ethics, one of which is to preserve the rights of others. People have fundamental rights, which others have a responsibility to defend, according to rights ethics.

4.8.3. Virtue Ethics

The goal of virtue ethics is to figure out what kind of person we should be. Moral differentiation and goodness are often used to describe virtue. A desirable and useful quality is exhibited by a virtuous person.

Actions are deemed right if they support positive character qualities (virtues such as responsibility, honesty, competence, loyalty, citizenship, trustworthiness, fairness, caring, and respect) and wrong if they support bad character traits (virtues such as selfishness, avarice, and greed) (vices like dishonesty, disloyalty, irresponsibility, or incompetence).

It is directly linked to personal character, with the mindset that "we do good things because we are virtuous individuals who desire to improve these character attributes in ourselves and others." This approach may appear to be primarily concerned with personal ethics, rather than engineering or professional ethics.

However, if a conduct is virtuous in a person's personal life, it is also virtuous in his or her work life. Virtue ethics is more abstract and resistant to detailed investigation. Nonhuman entities, such as corporations or governments, are more difficult to characterize in terms of virtue. Words that appear to be virtues on the surface but lead to vices might cause problems (like honor).

In engineering careers, one can use virtue ethics by answering questions like: Is this activity honest? Will this behavior show my commitment to my community and/or my employer?

Have I been accountable in my actions? Frequently, the answers to these questions reveal the best path of action. To apply virtue ethics to an ethical problem, one must first determine whether virtues or vices are appropriate for the situation. Then decide what each of these proposes as a path of action.

Companies, like individuals, should be held accountable for their ethical behavior, even if the ability to do so within the legal system is restricted. To put it another way, when it comes to an ethical issue, corporate misbehavior should not be masked behind a corporate mask.

Just because a company isn't a moral agency like a person doesn't imply it can do whatever it wants.

4.9. THE IMPORTANCE OF ETHICAL CONDUCT IN ENGINEERING

Some colleges require engineering ethics and professional behavior to be taught as part of the curriculum. In fact, most engineering programs require at least two credits of ethics education for engineering students. Furthermore, there are various reasons why engineering ethics is so important in engineering practice. Some of them include maintaining safety, honesty, and integrity.

4.9.1. Maintaining Public Safety

Engineers are accountable for preserving public safety, which is one of the main reasons why engineering ethics is so important. They may put people's lives at danger if they do not follow the technical code of ethics. Engineers who follow a professional code of ethics commit to prioritize the safety of society members when doing their duties. It suggests that engineers will use standard and approved materials throughout their careers, as well as standard engineering processes.

4.9.2. Integrity and Honesty

Engineers must uphold two of the most fundamental values: honesty and integrity. Engineers must be honest in all of their contacts with clients, employers, and the general public, according to the engineering code of professional ethics. Engineers must adhere to a code of ethics to ensure that they are truthful in all of their dealings (Figure 4.7).

Figure 4.7. Integrity and honesty play an important role.

Source: Image by Sound Teaching.

They must also uphold the engineering profession's ethics by refraining from engaging in any fraudulent or misleading practices. Engineers that follow the code of ethics will remain honest even when confronted with a difficulty or other distractions at work.

4.9.3. Promotes Public Confidence and Trust in the Profession

Engineering ethics codes of conduct should be recognized as an important part of the engineering field because they help to establish public trust in engineers by proving that they are ethical people who will do the right thing even if no one is looking.

It also ensures that people in adjacent businesses, such as construction, manufacturing, software development, and so on, are protected when it comes to safety and quality requirements.

Without these standards, each engineer would have to construct his or her own set of standards, which could lead to challenges like not knowing what constitutes acceptable behavior or how major design specifications should be determined without first contacting relevant stakeholders.

4.9.4. Protects Clients and Employers from Harm

Engineers are required by engineering standards of ethics to respect their clients' privacy and keep personal information private. They should not divulge any personal information about the client, including their name, age, gender, or location, as well as project details. Similarly, unless specifically authorized, the engineer should not reveal their employer's information.

4.9.5. Promotes Ethical Decision-Making in Circumstances of Uncertainty

Engineers, like everyone else, encounter problems, dilemmas, and moral dilemmas in their work. Engineers are governed by a code of ethics while making tough decisions, ensuring that they choose what is morally proper. To maintain moral ideals, they draw a clear line between what decisions are moral and promote social benefit rather than self-interest.

4.10. PARTS OF COMMUNICATION SYSTEM

- **Sender**: The Sender is the individual who transmits the communication. It may be a transmitting station from which the signal is broadcast.
- **Channel**: The channel is the medium through which communication signals travel to their intended destination.
- **Receiver**: The receiver is the individual who receives the message. It could be a receiving station that receives the signal sent out.

4.10.1. Signal

Signaling is the process of conveying information by gestures, noises, actions, and other means. As a result, a signal can be a source of energy that communicates data. This signal facilitates communication between the transmitter and the recipient. In communication systems, a signal is an electrical impulse or an electromagnetic wave that travels a distance to deliver a message.

4.11. TYPES OF SIGNAL

Signals are divided into two categories based on their characteristics: analog and digital. Analog and digital signals are further divided into categorics, as indicated in the diagram below.

4.11.1. Analog Signal

An analog signal is a continuous time-varying signal that reflects a time-varying quantity. This signal changes throughout time based on the current values of the quantity that represents it.

4.11.2. Digital Signal

A digital signal is defined as a signal that is discrete in nature or has a non-continuous shape. Individual values for this signal are denoted separately and are not reliant on past values, as if they were derived at that specific time. Digital values are usually defined as binary digits with only 1s and 0s. As a result, digital signals are also known as signals that represent 1s and 0s. Digital communication is defined as communication based on digital signals and digital values.

4.11.3. Periodic Signal

A Periodic Signal is any analog or digital signal that repeats its pattern over a period of time. This signal has a consistent pattern that is easy to guess or calculate.

4.11.4. Aperiodic Signal

Aperiodic Signal refers to any digital or analog signal that does not repeat its pattern over timc. This signal's pattern continues, but it is not repeated and is difficult to assume or quantify. In general, the signals used in communication systems are analog in nature, and depending on the demand, they are either conveyed in analog or transformed to digital and then communicated.

However, in order for a signal to be broadcast over a long distance without being affected by extraneous interferences or noise, it must go through a process known as modulation.

4.12. SOME MORAL ISSUES IN ENGINEERING

Whistle-blowing, loyalty, conflicts of interest, risk, and safety, and environmental care and sustainability are five ethical issues that have garnered a lot of attention in engineering ethics, especially when a professional approach is used. This list of ethical issues in engineering is by no means extensive, but it serves as an example because all of the issues are representative of the ethical issues that arise in engineering, even if most of them also occur in other professions.

4.12.1. Whistle-Blowing

Whistle-blowing can be defined as an employee making public certain abuses within an organization against his or her immediate superiors' wishes (or order), with the goal of resolving the abuses or telling the right authorities or the public about the abuses. Whistle-blowing can occur both inside and outside the company; for example, if a whistle-blower informs the company's upper management of certain abuses against his or her direct superior's wishes; and outside the company; in the latter case, it can address regulatory authorities, the media, or the general public.

Whistleblowing is not unique to engineering, but there are several reasons why it is relevant in engineering and has gotten a lot of attention, especially in the early days of engineering ethics. One explanation is that engineers may be aware of risks and negative repercussions of specific technologies or engineering projects that others are unaware of due to their specialized knowledge and skills (Figure 4.8).

Figure 4.8. Ethics issues in engineering.

Sources: Image by: blue diamond gallery.

Engineers are also frequently employed in hierarchical organizations (like companies). This means they may be forced to choose between serving their employer's interests and serving the public good. Although engineering societies' codes of ethics often emphasize the latter, their legal commitment is often to their company first and foremost. This may put them in a position where they must blow the whistle in the public good.

This scenario has sparked discussions about engineering ethics and efforts to strengthen the organizational and legal status of potential whistleblowers. Some businesses, for example, have implemented internal whistle-blowing procedures. Whistleblowers are protected by legislation in several countries.

Despite these efforts, whistleblowing often carries significant disadvantages for the whistleblower. The solution is likely to be found in measures that avoid whistle-blowing or make it a tactic to be employed only as a last resort, rather than in still better protection for whistle-blowers. Whistleblowing is not the best technique to deal with ethical difficulties in engineering in general.

It would be far preferable to work toward a scenario in which ethical concerns may be discussed openly and freely within organizations and with relevant stakeholders. This may necessitate more organizational and institutional reforms, as well as the ability for engineers to debate ethical concerns with management, clients, stakeholders, and the general public.

4.12.2. Loyalty

Engineers should be devoted to their employers and clients, according to engineering society codes of conduct. Engineers "must work for each employer or customer as faithful agents or trustees," according to the code of conduct of the National Society for Professional Engineers (NSPE) in the United States (NSPE 2007).

This loyalty may be incompatible with the duty to serve the public good. According to the NSPE code, the latter responsibility is more significant in specific situations. "Engineers shall not finalize, sign, or seal plans and/or specifications that are not in line with applicable engineering standards," it says, for example.

This article could be interpreted to mean that public commitments occasionally take precedence over the duty to be loyal to one's employer. However, one could argue that loyalty does not always mean complying with all of an employer's requests. Harris, Pritchard, and Rabins, for example, adopted the latter path in their book on engineering ethics.

They differentiate between what they refer to as uncritical and critical loyalty. They define unquestioning loyalty as "putting the employer's interests, as defined by the employer, ahead of any other concern" (Harris et al. 2005: 191). However, such blind loyalty may be wrong.

One might not only dispute on what the employer's interests are, allowing for some critical reflection, but one might also question whether the company's interests should always take precedence over other concerns, especially when the public is at risk. As a result, Harris, Pritchard, and Rabins suggest the concept of critical loyalty, which they define as "paying attention to the employer's interests, to the extent that this is practicable within the restrictions of the employee's personal and professional ethics" (Harris et al. 2005: 192).

4.12.3. Conflicts of Interest

A professional has a conflict of interest if pursuing that interest would conflict with meeting his or her professional obligations (including those to clients and employers) or impair his or her professional judgment. There are a few comments about this definition. To begin with, the presence of a conflict of interest does not entail real wrongdoing or failure to meet professional obligations.

Although apparent conflicts of interest do not necessarily imply moral wrongdoing, they can affect PEs' objectivity and trustworthiness because the professional is in a situation where his professional judgment could be harmed, even if this does not happen. As a result, obvious conflicts of interest should be avoided.

Second, not all cases of competing interests are conflicts of interest in the sense of the preceding definition. Rather, the phrase alludes to a potential impairment of one's judgment or a breach of one's professional responsibility. Third, an interest that may clash with professional judgment or obligations should be defined widely; it can be a professional or personal interest, and it can also include influences, loyalties, or temptations that are not strictly speaking interests but may impair professional judgment.

Bribery is an example of an obviously undesirable conflict of interest in engineering; however, there are also fewer clear-cut situations, such as taking presents. Conflicts of interest can also arise if company engineers serve on a standardization committee when the company has a vested interest in certain standards being accepted over others.

In general, it is preferable to prevent conflicts of interest, but this is not always achievable; for example, in standards committees, firm engineers may be useful due to their expertise and knowledge. If conflicts of interest are unavoidable, they should at the very least be disclosed to the appropriate parties.

Apart from avoiding conflicts of interest and disclosing them, there may be other ways to deal with them effectively, such as removing from the decision-making process or having an independent review of particular engineering decisions or judgments.

4.12.4. Safety and Risk

One of the key professional obligations of engineers is to ensure safety. Indeed, safety, and the protection of human health are top priorities in several engineering fields such as chemical engineering, mechanical engineering, and civil engineering, as well as biotechnology. Although safety is a top priority in engineering, it is not always clear how the concept of safety should be interpreted.

The absence of risk is one possible definition of safety. The downside of such a definition is that technical installations are never completely safe in the sense of being completely risk-free. Zero risk is impossible to achieve, and even if it could, it is typically undesirable because it will increase costs or compromise other technical principles such as sustainability (or privacy).

As a result, safety may be best characterized as the minimization of risks to the greatest extent practicable and morally desirable (given trade-offs with other values in engineering). Risk is commonly described in engineering as the product of the chance of an unwanted happening times the consequences of that event, while different definitions of risk can be found in the engineering literature as well (Hansson 2009).

Engineers frequently feel that the (moral) acceptability of hazards is proportional to their magnitude (defined as probability times consequences). However, the ethical literature on risk has indicated that a variety of other factors must be considered when determining if technology hazards are morally acceptable. The following factors have been identified as important for determining the moral acceptability of risk (e.g., Van de Poel and Royakkers 2011).

The first factor to evaluate is the risk-benefit ratio, which is primarily a utilitarian concern. A second consideration is whether people are willing to take risks and have given their informed permission to particular hazards. A third issue is the risk and reward distribution, and whether or not that distribution is fair or just.

A fourth concern is whether there are other technologies that can achieve the same result with less danger. Finally, whether those who cause or introduce the risk have good or evil intentions may be crucial, with possible intermediate examples of hazards caused by negligence or recklessness.

The contrast between safety and security threats may be crucial in this case. Security risks are caused by intentional harm, whereas safety risks are caused by unintended harm (such as natural causes or unintentional human error) (like terrorism, hacking or theft).

4.12.5. Environmental Care and Sustainability

While safety (and the preservation of human health) has long been recognized in engineering codes of ethics, environmental stewardship and sustainability are relatively new, at least in terms of codes of ethics. One factor could be that environmental concerns and sustainability emerged later in society than safety concerns (Figure 4.9).

Figure 4.9. Sustainability.

Sources: Image by Wallpaper Flare.

One could argue that they have gotten more social attention during the 1970s and 1980s, and that technical codes of ethics are simply lagging behind. In this context, it appears that, while engineers have always seen safety as an internal engineering objective, sustainability has long been regarded as a more political issue, and thus as more contentious.

However, this appears to be changing, since environmental care and sustainability are increasingly incorporated in engineering associations' codes of conduct and seen as fundamental engineering values. The relationship between engineering and the environment is unquestionably ambiguous.

In many ways, engineering and technology have been and continue to be sources of unsustainable development. At the same time, engineering and technology can help with environmental protection and sustainable development, and they may even be required to attain these goals.

The concept of sustainability can be understood and defined in a variety of ways, but the Brundtlandt definition of sustainable development is probably the most well-known: "Sustainable development is development that meets the needs of the present without compromising the ability of future generations to meet their own needs."

As this definition shows, the concept of sustainability encompasses more than environmental concerns; it also includes questions of social justice. In particular, sustainability can be argued to include the ideals of intergenerational justice (justice within the current generation) and intergenerational justice (justice beyond generations) (justice between generations).

As in the case of biofuels, the multiple value components of sustainability may in reality contradict with one another. Biofuels could be seen as a beneficial development in terms of intergenerational fairness, as they could help to assure the availability of fuels for future generations.

Their desirability is much more open to debate from an environmental or intragenerational standpoint, as their total environmental impact may be worse than traditional fuels, and by competing with food stocks, they may lead to rising food prices, negatively impacting food security in developing countries.

Environmental impact assessments, life cycle analysis, circular economy, and so-called design for sustainability and eco-design approaches are some of the tools and approaches that can be used to integrate sustainability and environmental considerations in engineering and technological development and design.

4.13. CONCLUSION

In the conclusion of this chapter, the importance of communication and ethics in the field of engineering is discussed. This chapter also discussed about the process of communication and the scope of engineering communication and ethics. In this chapter, professional codes of ethics as well as the professional approach to engineering ethics and codes of conduct have also been discussed.

Towards the end of the chapter, it discussed about the importance of ethical conduct in engineering such as maintaining public safety, integrity, and honesty, promotes public confidence and trust in the profession, promotes ethical decision – making in circumstances of uncertainty.

REFERENCES

1. Al-Naib, U., (2020). [online] Intechopen.com. Available at: https://www.intechopen.com/chapters/67276 (accessed on 1 July 2022).
2. Lal, R., (2019). [online] https://www.studocu.com/. Available at: https://www.studocu.com/in/document/delhi-technological-university/mechanical/introduction-to-engineering-materials-detailed-notes/14772746 (accessed on 1 July 2022).
3. Mukhtar, A., (n.d). *Introduction to Engineering Material and Their Applications*. [online] Academia.edu. Available at: https://www.academia.edu/6042047/Introduction_to_Engineering_Material_and_their_Applications (accessed on 1 July 2022).
4. Ronney, P. D., (n.d). *AME 101–Introduction to Mechanical Engineering and Graphics*. [e-Book] http://ronney.usc.edu/. Available at: http://ronney.usc.edu/AME101/AME101-LectureNotes.pdf (accessed on 1 July 2022).
5. Thorat, S., (n.d). [online] https://learnmech.com. Available at: https://learnmech.com/introduction-to-engineering-materia/ (accessed on 1 July 2022).

CHAPTER 5

ENGINEERING MATERIALS AND THEIR APPLICATIONS

CONTENTS

Engineering materials are used extensively in the design and manufacture of equipment and tools. Material selection for economy of operation and longevity of machinery. To be able to perform his duty reliably, an engineer must be familiar with the properties, uses, availability, and pricing of the materials used in construction/fabrication.

The subject of engineering material has been established to solve the aforementioned issues. Engineering material is described as: "A topic that deals with the manufacture, qualities, and applications of materials utilized in applied engineering."

5.1. INTRODUCTION

Engineering materials range in weight from lightweight to heavyweight. Alloys for aircraft, Semiconductor chips for computers, Photovoltaic for energy storage, Semiconductor scanners, and so on. Material means engineering materials, limited to solid materials only.

Science refers to the branch of applied science which deals with investigation of the relationship existing between the structure of materials and their properties. Materials differ from one another because of the difference in their properties for example, gold differs from iron because of its color, density, and corrosion resistance, among other things.

Property differences occur owing to variations in material structure. All solid materials are made up of a huge number of molecules that are linked together to create the bulk substance. Each molecule is made up of microscopic particles known as atoms. The qualities and structure of a material are determined by the individual properties of atoms and their order in the molecule.

A design engineer's understanding of materials and their characteristics is critical. The machine elements should be built of a material that is suitable for the operating circumstances. A design engineer must also be knowledgeable about the impact of manufacturing techniques and heat treatment on the characteristics of materials (Figure 5.1).

Figure 5.1. Image showing engineering material.

Source: Image by archdaily.com.

We will explore the most often used engineering materials and their qualities in this section. Metallurgy is the science and technique of economically extracting metals from their ores, purifying them, and preparing them for use. It investigates the microstructure of a metal, the structural details that may be observed under a microscope.

The microstructure of a metal influences its mechanical characteristics, particularly its elastic and plastic behavior when subjected to force. Chemical composition is the relative amount of a certain element in an alloy, generally given as a percentage of total weight. Microstructure will be determined by composition, as well as heat and mechanical treatments.

Metals and alloys are commonly utilized in everyday life. They are utilized to make machineries, bridges, motor vehicles, trains, building structures, ships, aircrafts, agricultural implements, and so on. As a result, increasing the quality and quantity of metal production in that country can result in real economic growth.

Most metals exist naturally in the combined state as minerals and are reactive. Only a few metals, such as gold, silver, platinum, and mercury, are found in their free form in the earth's crust. Metals with modest reaction rates exhibit minimal convergence to air, moisture, carbon dioxide, or non-metals present in nature.

A mineral is a naturally occurring material in which a metal or its compound exists. An ore is a mineral from which a metal may be profitably mined. Oxygen and carbon dioxide are the most active components found in nature, particularly in the atmosphere. Silicon and Sulphur may be found in abundance in the earth's crust.

Chloride ions are very abundant in seawater (obtained from dissolved salts). Because most active metals are electrically positive, they exist as distinct ions. As a result, the majority of the key ores of these metals exist as various components such as oxides, silicates, carbonates, and halides. Webster's dictionary defines materials as "substances from which anything is formed or created."

- Engineering Material: An inanimate matter component that is valuable to an engineer in the practice of his trade (used to produce products according to the needs and demands of society).
- Material Science: The study of the intrinsic structure, characteristics, and processing of materials, as well as their complex interactions/relationships.
- Material engineering is primarily concerned with the use of fundamental and applied understanding of materials in order for them to be turned into products as needed or desired by society (bridges materials knowledge from basic sciences to engineering disciplines).

5.2. IMPORTANCE OF ENGINEERING MATERIALS IN PRESENT WORLD

Materials are most likely more ingrained in our society than most of us think. Materials have an impact on almost every aspect of our everyday existence, including transportation, housing, clothes, communication, response, and food production.

Materials have helped to progress a variety of technologies, including medical and health, information, and communication, national security and space, transportation, structural materials, arts, and literature, textiles, personal hygiene, agriculture, and food science, and environmental science (Figure 5.2).

Figure 5.2. Non-ferrous metal working waste from Barton upon Humber.

Source: Image by Wikimedia commons.

The enthusiasm of Materials Science and Engineering is heightened by its close relationships with other disciplines and its influence on everyday life. Inter-disciplinary interactions between material sciences and other subjects in the creation of novel materials and their applications necessitate close contact and clear communication between scientists working in many domains.

As the contribution of materials science and engineering to other disciplines grows, scientists of all backgrounds will need to better understand how to engage in collaborative activities with other disciplines. Although it is not possible for scientists to master a broad corpus of scientific information across numerous fields, scientists must learn the abilities necessary to master certain areas.

5.3. THE EVOLUTION OF ENGINEERING MATERIALS

Materials have historically constrained design. The ages of man are called by the materials he used: stone, bronze, and iron. When he died, the materials he cherished were buried with him: Tutankhamen with shards of colored glass in his stone coffin, Agamemnon with his bronze sword and gilded mask, each signifying the top technology of his day.

What would they have taken with them if they had lived and died today? Perhaps their titanium watch; perhaps their carbon-fiber reinforced tennis racquet; perhaps their metal-matrix composite mountain bike; perhaps their polyether-ethyl-ketone crash helmet. This is not the era of a single substance; it is the age of a vast array of materials.

There has never been a time when the evolution of materials has been quicker or the spectrum of their characteristics has been more diverse. The list of materials available to engineers has grown so quickly that designers who graduated from college twenty years ago might be forgiven for not knowing half of them exist.

But, for the designer, not knowing is a recipe for catastrophe. Often, innovative design entails creatively using the features of new or better materials. And for the average person, even a schoolboy, not knowing is to lose out on one of our generation's most significant developments: the age of sophisticated materials.

Ceramics and glasses, natural polymers and composites were the materials of prehistory (> 10 000 BC, the Stone Age). Weapons, which were always at the pinnacle of technology, were fashioned of wood and flint, as were houses and bridges. Gold and silver were naturally occurring but had only a modest part in technology.

> The discovery of copper, bronze, and then iron (the Bronze Age, 4000 BC-1000 BC and the Iron Age, 1000 BC-AD 1620) stimulated enormous advances, replacing older wooden and stone weapons and tools (there is a cartoon on my office door, placed there by a student, depicting an irate Celt confronting a swordsmith with the words 'You sold me this bronze sword last week and now I'm supposed to upgrade').

Metals' supremacy in engineering was established by cast-iron technology in the 1620s, and the evolution of steels (1850 forward), light alloys (1940s), and special alloys since then have maintained their position. By the 1960s, 'engineering materials' had come to mean metals.' Engineers were given metallurgy classes; other materials were scarcely discussed (Figure 5.3).

Figure 5.3. Iron and steel scrap.

Source: Image by picryl.com.

Of course, there have been advancements in the other types of material. Ceramics included Portland cement, refractories, and fused silica, while polymers included rubber, Bakelite, and polyethylene, although their percentage of the total materials market was limited. Everything has changed since 1960.

The rate of development of new metallic alloys is presently modest; demand for steel and cast iron has actually declined in several nations. The polymer and composite industries, on the other hand, are quickly expanding, and estimates for the rise of production of new high-performance ceramics imply that this sector will also expand fast.

5.4. CURRENT TRENDS AND ADVANCES IN MATERIALS

Lumber, steel, and concrete are three materials that are consistently used in designing applications. Steel utilization in any nation is viewed as a proportion of its financial prosperity. Plan specialists keep on driving interest for different high steel composites for high-temperature applications like steam and gas turbines.

Amalgams including chromium, nickel, molybdenum, and tungsten, as well as iron, are more qualified for the previously mentioned purposes. Newer materials with combination resistance to high temperature and corrosion are fast becoming available, and material scientists and engineers are hard at work developing them.

Different types of ceramics, although being difficult to form and process, are finding usage at high temperatures. Innovative metallic materials that have recently been developed, in conjunction with new processing techniques like as isostatic pressing and isothermal forging, are capable of improving the fatigue characteristics of aviation components.

Powder metallurgy is capable of imparting enhanced mechanical characteristics under various loading circumstances while generating finished surfaces and lowering metal cutting costs.

Apparently, rapid cooling technology with cooling rates approaching one million degrees Celsius per second is being utilized to make metal powders that may be employed in product manufacturing processes such as powder metallurgy and hot isostatic pressing to obtain temperature resistant components.

To achieve anticorrosion qualities at high temperatures, metallurgists have developed various molybdenum and aluminum alloys, as well as titanium and nickel alloys. Polymeric materials are expanding at a 9% yearly pace and have increased in bulk faster than any other substance. Plastics have replaced metals, wood, glass, and paper in a variety of uses.

A recent trend in plastic technology is the creation of synergistic plastic alloys, which have superior qualities to the single members producing alloy parts. The recent discovery of plastic conductors may have a far-reaching influence in the near future. Ceramics are mostly used in high temperature, low load handling material.

Ceramics' main disadvantage is its brittleness and difficulty in cutting and shaping. Ceramics, when combined with metal powders such as molybdenum, generate cements, which are believed to be useful cutting materials. Cement tool bits are supposed to find a variety of applications in achieving fast cutting speeds and generating a higher surface polish.

Alumina, a well-known ceramic, is predicted to be effectively reinforced with molybdenum fibers. Attempts to improve the strength of such composite ceramics have so far been unsuccessful due to microcracking of molybdenum fibers. However, such composites have been proven to be more impact and thermal shock resistant.

5.4.1. Further Advances in Materials Development From View Point of New and Advanced Materials

Recent advancements, particularly in terms of new and advanced materials, can be classed as follows: Advanced Materials, Smart Materials, Nanostructured Materials/Nanotechnology, Quantum Dots (QDs), Spintronics, and Fermionic Condensate Matter.

5.4.2. Advanced Materials

Materials used in high-technology (or high-tech) applications are also referred to as advanced materials. We define high technology as a device or product that performs or functions on highly complicated and sophisticated principles, such as electronic equipment (VCRs, CD players, etc.), computers, fiber optic systems, spacecraft, airplanes, and military rocketry.

These advanced materials are usually either old materials with improved qualities or newly invented high-performance materials. Furthermore, advanced materials might be of any material type (e.g., metals, ceramics, or polymers) and are often rather costly.

Following chapters address the characteristics and uses of a variety of sophisticated materials, such as those used in lasers, integrated circuits (ICs), magnetic information storage, liquid crystal displays (LCDs), fiber optics, and the thermal projection system for the space shuttle orbiter.

5.4.3. Smart Materials (Materials of the Future)

Smart or intelligent materials are a class of innovative and cutting-edge materials that are currently being developed and will have a big impact on many of our technology. The term "smart" indicates that these materials can potentially detect changes in their environment and respond to these changes in predetermined ways—traits shared by live beings (Figure 5.4).

Figure 5.4. Corinthian pottery – ceramics.

Source: Image by Flickr.

Furthermore, the notion of smart materials is being expanded to quite intricate systems comprised of both smart and standard materials. The area of smart materials intends to connect the sensor (which detects an input signal), actuator (that further performs a responsive and adaptive function), and control circuit it in to a single integrated unit.

Actuators should be used to modify the form, location, natural frequency, or mechanical performance in response to changes in temperature, electric fields, and/or magnetic fields.

5.5. CLASSIFICATION OF ENGINEERING MATERIAL

The factors which form the basis of various systems of classifications of materials in material science and engineering are:

- The chemical composition of the material;
- The mode of occurrence of the material in nature; and

- The refining and manufacturing process to which the material is subjected prior to acquiring the required properties, The material's atomic and crystalline structure, as well as
- Its industrial and technological applications.

Common engineering materials that are within the purview of material science and engineering can be divided into the following categories:

- Metals (ferrous and non-ferrous) and alloys;
- Ceramics;
- Organic Polymers;
- Composites;
- Semi-conductors; and
- Biomaterials.

5.5.1. Metallic Materials Are Typically Classified According to Their Use In Engineering As Under

- **Pure Metals:** In general, obtaining pure metal is quite difficult. They are usually obtained by processing the ore. Pure metals are often useless to engineers. However, using specialized and expensive processes, pure metals (purity 99.99%), such as aluminum and copper, may be obtained.
- **Alloyed Metals:** Alloys are formed by combining two or more metals, at least one of which is metal. The qualities of an alloy can be very different from the component substances, for example, 18–8 stainless steel includes 18% chromium and 8% nickel, but low carbon steel contains less than 0.15% carbon and is exceptionally tough, extremely ductile, and highly resistant to corrosion. It is important to notice that these qualities differ significantly from the behavior of original carbon steel.
- **Ferrous Metals:** These ferrous metals are mostly composed of iron. Nonferrous metals are present in large amounts in ferrous alloys. Ferrous alloys are particularly significant in engineering. These can be categorized into the following groups based on the proportion of carbon and alloying elements present:
- **Dead Mild Steel:** Carbon content in these materials can reach 0.15%. These are quite ductile and not particularly strong. These materials have a cheap manufacturing cost.

- **Mild Steels:** Carbon percentages in these materials range from 0.15% to 0.45%. These are reasonably robust and easily welded. These materials are also inexpensive to produce.
- **Medium Carbon Steels:** Carbon content ranges from 0.45% to 0.6%. The strength of these materials is strong, but their weldability is low.
- **High Carbon Steels:** These contain carbon in concentrations ranging from 0.6 to 1.5%. Heat treatment hardens and toughens these materials, although weldability is low. Plain carbon steel is steel that has a carbon concentration of up to 1.5%, a silica content of up to 0.5%, and a manganese level of up to 1.5%, together with traces of other components.
- **Cast Iron:** The carbon content of these compounds ranges from 2% to 4.5%. These compounds are produced at a minimal cost and are utilized as ferrous casting alloys.
- **Non – Ferrous Metals:** These compounds are made up of metals other than iron. However, they may contain trace amounts of iron. Only seven non-ferrous metals are accessible in sufficient quantity and at a reasonable cost to be employed as common engineering metals.

These are aluminum, tin, copper, nickel, zinc, magnesium. Other nonferrous metals, around fourteen in number, are produced in very modest amounts yet are critical in modern industry. Chromium, mercury, cobalt, tungsten, vanadium, molybdenum, antimony, cadmium, zirconium, beryllium, niobium, titanium, tantalum, and manganese are among them.

- **Sintered Metals:** When compared to the metals from which these things were cast, these compounds have significantly different characteristics and structures. Sintered metals are created using the powder metallurgy technology. The metals to be sintered are first powered and then combined in the proper estimated quantities.

After correctly mixing, they are placed in the required form die and treated under pressure. Finally, they are sintered in a furnace. It should be noted that the resulting combination is not a real alloy, although it does have some of the qualities of normal alloys.

- **Clad Metals:** To take use of the qualities of both materials, a' sandwich' of two materials is constructed. Cladding is the word for this procedure. By rolling the two metals together when they are red hot, stainless steel is primarily encrusted with a thick coating of mild steel.

This approach prevents rusting on a single surface. Another use for this method is the cladding of dualism with thin sheets of pure aluminum. The surface layers of aluminum, i.e. the exterior layers, resist corrosion, whilst the inside layer of duralumin provides exceptional strength. This approach is quite inexpensive to produce.

The selection of a proper material, for engineering purposes, is one of the most difficult problems. The best material is one which serve the desired objective at the minimum cost. One of the most challenging tasks of an engineer is the proper selection of the material for a particular job/ component of a machine or structure. An engineer must be in a position to choose the optimum combination of properties in a material at the lowest possible cost without compromising the quality (Figure 5.5).

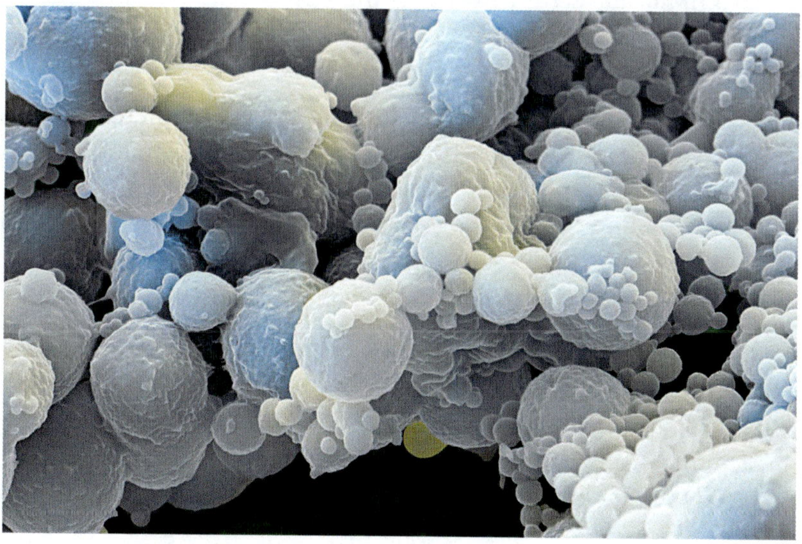

Figure 5.5. Biomaterials made of natural self-assembling proteins.

Source: Image by Flickr.

The properties and behavior of a material depends upon the several factors, e.g., composition, crystal structure, conditions during service and the interaction among them. The performance of materials may be found satisfactory within certain limitations or conditions. However, beyond these conditions, the performance of materials may not be found satisfactory.

The Major Factors Affecting the Selection of Materials Are:

- Component shape;
- Dimensional tolerance;
- Mechanical properties;
- Fabrication requirements;
- Service requirements;
- Cost of the material;
- Cost of processing; and
- Availability of the material.

All of these primary considerations have a complicated impact on material selection.

Obviously, the material selection becomes restricted; for example, one can only use materials with lower melting points, such as aluminum, zinc, magnesium, and thermoplastics. Some materials can be polished to a high degree of precision, whereas others cannot.

Clearly, the needed dimensional tolerance for completed components will impact material selection. All mechanical attributes, such as hardness, strength, and so on, aid us in selecting a suitable material for given situations. The method of material processing also influences the qualities of a component; for example, forged components can be stronger than casted components.

Different sorts of working methods might result in different types of fiber structure. Investment casting, on the other hand, may deliver exact dimensions at a lower cost than machine operations. Dimensional stability, strength, toughness, heat resistance, corrosion resistance, fatigue, and creep resistance, electrical, and thermal conductivity, and so on are all service criteria.

Fabrication criteria include cast ability, or ease of casting a material, weldability, or ease of welding a material, machinability, or ease of machining a material, formability, or ease of forming a material, hardenability, and so on.

In most situations, the cost of raw materials accounts for around half of the total cost. Clearly, the cost of the material is an important aspect that determines the material or process selection.

5.5.2. Mechanical Properties of Metals

The mechanical characteristics of metals are those linked with the materials' capacity to withstand mechanical forces and loads. Strength, stiffness, elasticity, plasticity, ductility, brittleness, malleability, toughness, resilience, creep, and hardness are some of the mechanical characteristics of metal. We will now go over these things in detail:

- **Strength:** It is a material's capacity to withstand external pressures without breaking.
- **Stiffness:** It is a material's capacity to withstand deformation under stress. The modulus of elasticity is a stiffness measurement.
- **Elasticity:** A material's ability to recover its original shape after deformation when external pressures are eliminated. This feature makes materials used in tools and machinery preferable. It should have been noted that steel is more flexible than rubber.
- **Plasticity:** It is a material quality that permanently preserves the deformation caused by load. This material quality is required for forgings, impressing pictures on coins, and decorative work.
- **Ductility:** It is the property of a substance that allows it to be dragged into wire by applying tensile force. A ductile material must be strong as well as pliable. The words percentage elongation and% decrease in area are commonly used to describe ductility.

In order of decreasing ductility, the ductile materials widely utilized in engineering practice include mild steel, copper, aluminum, nickel, zinc, tin, and lead. Note: The ductility of a material is frequently assessed in tensile tests using% elongation and percentage decrease in area.

- **Brittleness:** It is the antithesis of ductility in a substance. It is the property of a substance to break with little lasting deformation. When exposed to tensile stresses, brittle materials snap off with no discernible elongation. Cast iron is a fragile substance.
- **Malleability:** It is a type of ductility that allows materials to be rolled or pounded into thin sheets. A malleable substance should be plastic, but it is not required to be extremely strong. Lead, soft steel, wrought iron, copper, and aluminum are the malleable materials typically utilized in engineering practice (in decreasing order of malleability).

- **Toughness:** It is a material's ability to withstand fracture caused by strong impact stresses such as hammer strikes. When heated, the material's hardness lowers. It is determined by the amount of energy absorbed by a unit volume of the material after being strained to the point of fracture. This feature is desirable in portions that will be subjected to shock and impact loads.
- **Machinability:** It is a material attribute that relates to a relative scenario in which a material can be sliced. A material's machinability may be determined in a variety of ways, such as comparing the tool life for cutting different materials or the thrust required to remove the material at a certain pace or the amount of energy necessary to remove a unit volume of material It should be noted that brass is easier to process than steel.
- **Resilience:** It is a material's ability to absorb energy while also resisting shock and impact loads. The quantity of energy absorbed per unit volume inside the elastic limit is used to calculate it. This characteristic is critical for spring materials.
- **Creep:** When a part is subjected to sustained stress at high temperatures over an extended length of time, it will suffer creep, which is a gradual and irreversible deformation. This feature is taken into account while developing internal combustion engines, boilers, and turbines.
- **Tiredness:** When a material is repeatedly stressed, it fails at stresses lower than the yield point stresses. Fatigue is a term used to describe this sort of material breakdown. The failure is caused by a gradual fracture development, which is typically tiny and microscopic in size. This feature is taken into account while constructing shafts, connecting rods, springs, and gears, among other things.
- **Hardness:** It is an extremely significant feature of metals and has a wide range of implications. It encompasses a wide range of qualities including as resistance to wear, scratching, distortion, and machinability.

It also refers to a metal's capacity to cut another metal. The hardness is commonly stated in numbers that vary depending on how the test is performed. The following tests can be used to determine a metal's hardness:

- Brinell hardness test;
- Rockwell hardness test;
- Vickers hardness (also called Diamond Pyramid) test; and
- Shore ceroscopy

5.6. APPLICATIONS OF ENGINEERING MATERIALS

5.6.1. Structural Applications

Structural applications demand mechanical performance (strength, stiffness, and vibration damping ability) from the material, which may or may not carry the load in the structure. Mechanical property criteria are very stringent when the material carries the load.

A building with steel-reinforced concrete columns bearing the weight of the structure and unreinforced concrete architectural panels covering the front of the building is an example. Although both the columns and the panels have structural purposes and are structural materials, only the columns bear weight.

The panels must have mechanical strength and rigidity, but the columns must meet much higher standards. Buildings, bridges, piers, highways, landfill covers, airplanes, automobiles (body, bumper, drive shaft, window, engine components, and brakes), bicycles, wheelchairs, ships, submarines, machinery, satellites, missiles, tennis rackets, fishing rods, skis, pressure vessels, cargo containers, furniture, pipelines, utility poles, armored vehicles, utensils, fasteners, and so on are examples of structures.

A structural material may be required to have other features in addition to mechanical properties, such as low density (lightweight) for fuel savings in airplanes and automobiles, rapid speed in racing bicycles, and handleability in wheelchairs and armored vehicles.

Another characteristic that is frequently required is corrosion resistance, which is important for the longevity of all constructions, notably autos and bridges. Another quality that may be required is the capacity to tolerate high temperatures and/or thermal cycling, as the structure may be exposed to heat during operation, maintenance, or repair.

The ability of a structural material to perform tasks other than structural ones is a relatively new trend. The material becomes multipurpose, which reduces costs and simplifies design. Damage detection is an example of a non–structural function. Such sensing, also known as structural health monitoring, is useful for danger prevention. It is especially critical for aged airplanes and bridges.

The sensing function can be achieved by embedding sensors in the structure (such as optical fibers, the damage or strain of which influences light throughput). However, embedding frequently results in mechanical property deterioration, in addition, embedded gadgets are more expensive and less durable than structural materials.

Another technique to achieve the sensing function is to detect changes in the structural material's property (e.g., electrical resistivity) caused by damage. As a result, the structural material acts as its own sensor and is referred to as "self-sensing."

5.6.2. Electronic Applications

Because the electrical, optical, and magnetic characteristics of materials are mainly regulated by electrons, electronic applications encompass electrical, optical, and magnetic applications. These three areas of application have some overlap.

Electrical applications include computers, electronics, electrical circuitry (resistors, capacitors, and inductors), electronic devices (diodes and transistors), optoelectronic devices (solar cells, light sensors, and light-emitting diodes for the conversion of electrical energy to optical energy), thermoelectric devices (heaters, coolers, and thermocouples for the conversion of electrical energy to thermal energy),

and piezoelectric devices (strain sensors and actuators), micromachines (or microelectromechanical systems, MEMS), ferroelectric computer memories, electrical interconnections (solder joints, thick-film conductors, and thin-film conductors), dielectrics (electrical insulators in bulk, thick-film, and thin-film forms), heat sinks, electromagnetic interference (EMI) shielding, cables, connectors, power supplies, electrical energy storage, motors, electrical contacts, and brushes (the use of a magnetically induced electrical current to indicate flaws in a material).

Lasers, light sources, optical fibers (materials with low optical absorptivity for communication and sensing), absorbers, reflectors, and transmitters of electromagnetic radiation of various wavelengths (for optical filters, low-observable or Stealth aircraft, randoms, transparencies, optical lenses, and so on), photography, photocopying, optical data storage, holography, and color control are all examples of optical applications.

Transformers, magnetic recording, magnetic computer memories, magnetic field sensors, magnetic shielding, magnetically levitated trains, robotics, micromachines, magnetic particle inspection (the use of magnetic particles to indicate the location of flaws in a magnetic material), magnetic energy storage, magnetostriction (strain in a material due to the application of a magnetic field),

Magnetorheological fluids (for magnetic field-controlled vibration damping), magnetic resonance imaging (MRI, for patient diagnosis in hospitals), and mass spectrometry (for chemical analysis).

5.6.3. Thermal Applications

Thermal applications involve heat transmission, whether by conduction, convection, or radiation. Heat transfer is required in building heating, industrial processes such as casting and annealing, cooking, de-icing, and so on.

As well as cooling of buildings, refrigeration of food and industrial materials, cooling of electronics, removal of heat generated by chemical reactions such as cement hydration, removal of heat generated by friction or abrasion as in a brake and as in machining, removal of heat generated by electromagnetic radiation impingement, heat removal from industrial operations such as welding, and so on.

Conduction is the transfer of heat in a substance from points of higher to locations of lower temperature. Metals are commonly used because of their excellent heat conductivity. The flow of a heated fluid causes convection. Pushed convection occurs when the fluid is forced to move by a pump or a blower. Natural or free convection occurs when the fluid flows owing to changes in density.

The fluid can be either a liquid (oil) or a gas (air), and it must be able to tolerate the heat. Fluids are not covered in this text. Space heaters use radiation, namely blackbody radiation. It refers to the body's continuous release of radiant radiation. The energy is often in the form of infrared electromagnetic radiation.

The dominating wavelength of the emitted radiation decreases as the body temperature rises. The higher the temperature, the greater the rate of radiant radiation output per unit area of the surface. T4 is the absolute temperature, and this rate is proportional to it. It is also proportional to the body's emissivity, which varies according to the substance. It rises in particular with increasing surface roughness.

5.6.4. Electrochemical Applications

Electrochemical applications are those that deal with electrochemical processes. An electrochemical reaction consists of an oxidation reaction (such as Fe Fe^{2+} + 2e–) that generates electrons and a reduction reaction (such as O_2 + $2H_2O$ + $4e^-$ $4OH^-$) that consumes electrons. The anode is the electrode that emits electrons; the cathode is the electrode that receives electrons.

Electrons flow from the anode to the cathode when the anode and cathode are electrically coupled. Anode and cathode must both be electronic conductors. Ions migrate in an ionic conductor as electrons flow in the wire from the anode to the cathode (referred to as the electrolyte) put between the anode and the cathode so that cations (positive ions) produced by anode oxidation flow in the electrolyte from the anode to the cathode.

The ability of an electrode to oxidize determines whether it acts as an anode or a cathode. The anode is the electrode with the greater tendency, whereas the cathode is the other electrode. A voltage, on the other hand, can be placed between the anode and the cathode at the position of the wire, with the positive end of the voltage at the anode.

The positive end draws electrons, pushing the anode to be oxidized, even though it is not more susceptible to oxidation than the cathode. as well as the cathode the anode corrodes as a result of the oxidation process. For example, the oxidation process Fe Fe^{2+} + 2e– corrodes iron atoms, resulting in Fe^{2+} ions that enter the electrolyte.

Corrosion prevention is achieved by impeding the oxidation reaction. Electrochemical reactions are important not just for corrosion, but also for batteries, fuel cells, and industrial operations that involve electrochemical reactions (such as the reduction of Al_2O_3 to generate Al). The combustion of fossil fuels such as coal and gasoline pollutes the environment.

Batteries and fuel cells, on the other hand, generate fewer environmental issues. A battery has an anode and a cathode, which have differing oxidation propensities. When the anode and cathode are open-circuited at the wire, there is a voltage differential between them, with the negative end of the voltage at the anode.

This is due to the anode's desire to emit electrons, but the electrons are unable to escape due to the open circuit state. This voltage differential represents the output of the battery, which is a direct current (DC).

5.6.5. Environmental Applications

Environmental applications are those that are concerned with safeguarding the environment from contamination. Protection might include either the elimination of a pollutant or a reduction in the number of pollutants produced. Pollutants can be removed through adsorption on the surface of a solid (e.g., activated carbon) with surface porosity.

It can also be accomplished by planting plants that absorb CO_2 gas. Pollutant generation can be reduced by changing the materials and/or processes used in industry, such as using biodegradable materials (materials that can be degraded by Nature, eliminating the need for disposal), recycling materials, or switching from fossil fuels to batteries, fuel cells, solar cells, and/or hydrogen.

Materials have been produced primarily for structural, electrical, thermal, or other uses, with little regard for disposal or recycling issues. It is widely acknowledged that such concerns must be made throughout the design and development of materials, rather than afterwards. Adsorption materials are critical in the development of materials for environmental applications.

Carbons, zeolites, aerogels, and other porous materials are among them. Large adsorption capacity, pore size large enough for relatively large molecules and ions to lodge in, ability to be regenerated or cleaned after use, fluid dynamics for fast movement of the fluid from which the pollutant is to be removed, and, in some cases, selective adsorption of specific species all seem to be desirable characteristics.

Due to the sheer channels between the fibers, activated carbon fibers outperform activated carbon particles in fluid dynamics. They are, however, far more costly. To act as adsorption sites, pores on the surface of a substance must be accessible from the outside. In general, pores can be classified as macropores (more than 500), mesopores (between 20 and 500), micropores (between 8 and 20), or micro-micropores (less than 8).

Micropores and micro-micropores are common features of activated carbons. Electronic pollution has emerged as a significant environmental issue. It is caused by electromagnetic waves (especially radio waves) existing in the environment as a result of radiation sources such as cellular devices.

Such radiation can interact with digital gadgets like computers, posing risks and disrupting society's operations. Radiation sources and electronics are protected by materials that reflect and/or absorb radiation to relieve this problem.

5.6.6. Biomedical Applications

Biomedical applications are concerned with the diagnosis, treatment, and prevention of illnesses, diseases, and impairments.

Implants (hips, heart valves, skin, and teeth), surgical, and diagnostic devices, pacemakers (devices for electrical control of heartbeats), electrodes for collecting or transmitting electrical or optical signals for diagnosis or treatment, wheelchairs, assistive devices, exercise equipment, pharmaceutical packaging (for controlled release of a drug into the body or for other purposes), and instrumentation for diagnosis and chemical analysis are among them (such as equipment for analyzing blood and urine).

Implants are particularly difficult; they must be built of biocompatible (compatible with fluids such as blood), corrosion-resistant, wear-resistant, and fatigue-resistant materials that can maintain these features over time.

5.7. THE FUTURE ENGINEERING MATERIALS

How will we deal with future shortages of engineering materials? The following are some obvious strategies:

5.7.1. Material-Efficient Design

Many present designs use more material than is required, or even use potentially rare resources where more available ones might suffice. When a surface feature (e.g., low friction or high corrosion resistance) is desired, a thin surface layer of the rare material bonded to a cheap abundant substrate can substitute the bulk use of a scarcer material.

5.7.2. Substitution

The designer is interested in the property rather than the material itself. Sometimes a more widely accessible material might be used to replace a limited one, albeit this typically requires a significant investment (new processing methods, new joining methods, etc.).

Substitution examples include the replacement of stone and wood in building with steel and concrete; the replacement of copper in plumbing with polymers; the transition from wood and metals to polymers in home products; and the transition from copper to aluminum in electrical wiring.

However, the replacement has technical limits. Some materials are utilized in ways that others cannot easily fill. Platinum as a catalyst, liquid helium as a refrigerant, and silver on electrical contact regions are irreplaceable; each serves a distinct purpose.

Others, such as a substitute for tungsten lamp filaments, are now undergoing extensive research. Finally, substitution raises demand for the substitute material, which may be in short supply. The widespread substitution of plastics for other materials places a greater strain on petrochemicals, which are currently sourced from oi.

5.7.3. Recycling

Recycling is not a new concept; for millennia, old construction materials have been recycled, and scrap metal has been recycled for generations; both are huge industry. Recycling is labor expensive, which is why increasing its reach is difficult.

5.8. MATERIALS AND THE ENVIRONMENT: GREEN DESIGN

Technical advancement and environmental responsibility are not mutually exclusive aims. Many civilizations throughout history have embraced environmentally sensitive lifestyles while making technical and societal advancements.

However, the acceleration of industrial expansion since the beginning of the industrial revolution has swamped the environment, with local and global effects that cannot be ignored. There is an increasing push to lessen and rectify this environmental damage.

It necessitates less harmful procedures and products that are lighter, less energy-intensive, and easier to recycle; all while maintaining product quality. New technologics that allow for increased output while having a lower environmental effect must (and can) be developed.

Concern for the environment must be incorporated into the design process, or brought "behind the drawing board," by considering a life-cycle view of the product that encompasses manufacturing, distribution, usage, and final disposal.

5.8.1. Energy-Content As a Measure of Environmental Impact

Every substance contains energy. Metals are mined, refined, and shaped with energy; ceramics and cement are fired with energy, and oil-based polymers and elastomers include energy. When you utilize a substance, you are utilizing energy, and energy has an environmental cost: CO_2, nitrogen oxides, Sulphur compounds, dust, and waste heat.

Energy is only one of the environmental effects of material production and usage, but it is the easiest to assess. The majority of polymers are generated from oil. As a result, they are said to be energy-intensive, which has consequences for their future.

The two graphs demonstrate that most polymers are less energy-intensive than primary aluminum, magnesium, or titanium per unit of function in bending (the most frequent mode of loading) and that some are comparable with steel.

The majority of the energy needed in the creation of light alloys such as aluminum and magnesium is used to convert the ore to the elemental metal, hence these materials are substantially less energy demanding when recycled. Efficient collection and recycling contribute significantly to energy savings.

5.8.2. The Pressure to Recycle and Reuse

There are several reasons why you should not toss things away. Discarded items harm the environment and constitute pollution. Materials that are withdrawn from the manufacturing cycle must be replenished using a natural resource. And materials have energy, which is lost when they are discarded. Recycling is clearly desired.

However, in a market system, it will only occur if there is a profit to be earned. First, consider where recycling works effectively and where it doesn't. Primary scrap – the byproducts of manufacturing such as turnings, trimmings, and tailings – has a high value and is almost entirely recycled. This is due to the fact that it is not tainted and is not distributed.

Secondary scrap has gone through a consumption cycle; the paper of a newspaper, the aluminum of a drink can, and the steel of a car are all tainted by other materials to which they are linked; corrosion products; ink and paint. And they are disseminated, some of them quite broadly dispersed, such as tungsten in lamp filaments.

They are worth nothing or less than nothing in this form, that is, the cost of collection exceeds the value of the junk itself. Nonetheless, this is by far the most significant component of the material cycle. Recycling newsprint and bottles, for example, is not economically feasible in a free market.

Recycling occurs, but it is dependent on social consciousness and goodwill, as well as municipal subsidies and awareness. It is dangerous for these reasons. Two things have the potential to transform everything. The most obvious example is legislation (a divergence from a fully free market economy).

A deposit or 'dispersal fee' embedded into the price of each product transforms the economics and efficacy of recycling; various civilizations have tried it, and it works. The other aspect is designed. The major challenges in recycling are identification, separation, and purification; all of these issues may be addressed by the designer.

Fingerprinting goods by color, symbol, or bar code provides for identification. Economic separation is enabled by the design for disassembly and the avoidance of mutually contaminated combinations. Decontamination is aided by clever chemistry (strippable paints; soluble glues).

Finally, design to avoid the need to recycle: longer primary life; and more consideration of secondary usage during the early design stage.

5.9. INTRODUCTION TO MATERIALS SELECTION

One of the most difficult tasks for a materials engineer is selecting the right material for the purpose, such as a specific component of a machine or construction. An engineer must be able to select the best mix of qualities in a material at the lowest feasible cost without sacrificing quality.

5.9.1. Factors Affecting the Selection of Materials

- **Component Shape:** The shape and size of a component have a significant impact on the selection of the processing unit, which in turn influences the selection of the material. To illustrate, suppose the best viable manufacturing process, under given conditions, is die casting.

Obviously, the material selection becomes restricted, i.e. one can only use materials with lower melting points, such as aluminum, zinc, magnesium, and thermoplastics.

- **Dimensional Tolerance:** Some materials can be finished to near tolerances, whereas others cannot. Clearly, the required dimensional tolerance for completed components will impact material selection.
- **Mechanical Properties:** All mechanical attributes, such as toughness, hardness, strength, and so on, aid us in selecting a suitable material for given situations.
- **Fabrication (Manufacturing) Requirements:** The method of material processing also influences the qualities of a component; for example, forged components can be stronger than casted components. Distinct working techniques might result in different types of fiber structures.

Investment casting, on the other hand, may deliver exact dimensions at a lower cost than machine operations. Fabrication criteria include: cast ability (ease of casting a material), weldability (easy of welding a material), machinability (ease of machining a material), formability (ease of forming a material), hardenability, and so on.

- **Service Requirements:** Service specifications are:
  - Dimensional Stability;
  - Strength;
  - Toughness;
  - Heat Resistance;
  - Corrosion Resistance;
  - Fatigue and Creep Resistance; and
  - Electrical and Thermal Conductivity etc.
  - Cost:
- **Cost of the Material:** In most situations, the cost of raw materials accounts for around half of the total cost. Obviously, the cost of the material is a crucial aspect that determines the material or process selection.

It should be noted that using less expensive materials does not necessarily result in a lower ultimate cost of the component or product. The use of less expensive materials may result in greater processing costs due to the vast number of procedures that must be done, as well as more scrap.

We can clearly understand how this sometimes causes the overall cost to be higher than that of expensive raw material combined with low processing costs owing to fewer operations and less waste. The material used influences the finer points of design.

As a result, the material as well as the procedure are chosen early in the design process; for example, whether the material is to be bonded by spot welding, screws, or rivets must be determined early in the design process.

- **Cost of Processing:** In most sectors, processing costs (labor costs) and other costs such as overhead charges account for around half of the total production cost. The overhead cost in the automation industries is significantly higher than the other expenditures. If all of these costs can be reduced, the overall manufacturing costs will decrease.
- **Availability of the Material:** We may discover that the availability of the material becomes a deciding factor at times. When the desired material supply is restricted, a pricey material that is abundantly accessible may be chosen.

5.9.2. Procedure for Materials Selection

- Design;
- Material selection;
- Process selection;
- Production;
- Evaluation; and
- Possible redesign or modification.

5.9.3. Modern Materials Needs and Challenges

Despite considerable advances in the field of material science in recent years, there are still technological hurdles, such as the creation of more sophisticated and specialized materials, as well as the influence of materials manufacturing on the environment.

Nuclear (fission and fusion) energy has considerable promise, but the solutions to the numerous outstanding issues will inevitably require materials ranging from fuels to containment buildings to radioactive waste disposal facilities. Fusion science has advanced at an astonishing rate in recent years.

Fusion materials and technology have advanced significantly, with prototypes of essential components of a fusion power plant created and successfully tested. Obviously, new high strength, low density structural materials are still being developed, as well as materials with improved temperature tolerance for usage in engine components.

Furthermore, there is an urgent need to discover new, cost-effective energy sources and to make better use of existing energy resources. Hydrogen appears to be the future fuel. Hydrogen has the most potential for environmental and energy supply improvements.

Hydrogen, like electricity, is a flexible energy carrier that may be produced from a wide range of readily available primary energy sources such as natural gas, coal, biomass, wastes, sunshine, wind, and nuclear power. Although hydrogen manufacturing processes exist, more improvement is desired for usage in zero-carbon energy systems.

Materials will surely play an important part in these changes. We know that the quality of the environment is dependent on our capacity to regulate air and water pollution. Various materials are used in pollution control procedures. Material processing and refining procedures must be improved to create less environmental harm (Figure 5.6).

Figure 5.6. Materials needs and opportunities.

Source: Image by serc.carleton.edu.

Toxic compounds are formed during the manufacturing procedures of particular materials; thus we must consider the environmental impact of their disposal. Many of the materials we use are produced from non-renewable resources, such as polymers, whose primary raw ingredient is oil, and some metals.

These non-renewable resources are progressively depleting, necessitating: Additional reserves are being sought. The creation of novel materials with equivalent qualities but a lower environmental effect. increased recycling efforts and innovative recycling technology.

5.10. CONCLUSION

In the conclusion of this chapter, it discussed about the engineering materials and their applications. It also discussed about the importance of the engineering materials in the present world. In this chapter, the evolution of the engineering materials has been discussed. Towards the end of the chapter, it discussed about the current trends and advances in the materials, various types of engineering material, and applications of engineering materials.

REFERENCES

1. (n.d). *Mechanical Properties of Engineering Materials*. [e-Book] https://www.bu.edu.eg. Available at: https://www.bu.edu.eg/portal/uploads/Engineering,%20Shoubra/Civil%20Engineering/2469/crs-6270/Files/MECHANICAL%20PROPERTIES.pdf (accessed on 1 July 2022).
2. Ashby, M., (1999). *Materials Selection Mechanical Design* (2nd edn.) [e-Book] http://www.utc.fr/. Available at: http://www.utc.fr/~hagegebe/UV/MQ12/CORRECTIONS_TD/%5BASHBY99%5D%20-%20Materials%20Selection%20In%20Mechanical%20Design%202Ed.pdf (accessed on 1 July 2022).
3. Chung, D., (2001). *Applied Materials Science*. [online] Available at: https://www.researchgate.net/publication/31715409_Applied_Materials_Science_Applications_of_Engineering_Materials_in_Structural_Electronics_Thermal_and_Other_Industries_DDL_Chung/link/02e7e52cb402f86116000000/download (accessed on 1 July 2022).
4. Mukhtar, A., (n.d). *Introduction to Engineering Material and their Applications*. [online] Academia.edu. Available at: https://www.academia.edu/6042047/Introduction_to_Engineering_Material_and_their_Applications (accessed on 1 July 2022).
5. Pandey, S., & Singh, V., (2015). *The Importance of Engineering Materials in Present World*. [e-Book] https://www.ijsr.net/. Available at: https://www.ijsr.net/archive/v6i3/ART20171428.pdf (accessed on 1 July 2022).

CHAPTER 6

MATHEMATICS, PROBABILITY, AND STATISTICS IN ENGINEERING

CONTENTS

Almost all engineering and applied science undergraduate programs now include at least one fundamental course on probability and statistical inference. The acknowledgement of the necessity to introduce probability theory concepts into a wide range of scientific domains today reflects some of the major shifts in science and engineering education over the last 25 years.

One of the most noteworthy changes is the increased emphasis on intricacy and precision. A scientist now understands the value of investigating scientific phenomena with intricate interrelationships among their components; these components are frequently not only mechanical or electrical in character, but also 'soft-science' in nature, such as those derived from behavioral and social sciences.

6.1. INTRODUCTION

Designing a comprehensive transportation system necessitates a thorough grasp of the problem's technological components as well as user behavior patterns, land-use restrictions, environmental needs, pricing strategies, and so on. Precision is also emphasized, both in defining the interrelationships among the components involved in a scientific phenomenon and in anticipating its behavior.

This, combined with the rising complexity of the challenges that is confront, leads to the awareness that a large deal of uncertainty and variability are invariably included in problem formulation, and probability and statistics are one of the mathematical instruments that can help us cope with them.

Probabilistic concepts are employed in a wide range of randomness – related scientific activities. Randomness is an empirical phenomena defined by the fact that the numbers we're interested in don't have a predictable outcome under a particular set of conditions, but instead have a statistical regularity linked with various probable outcomes.

The ratio m/n, where m is the number of observed instances of a certain result, tends to a unique limit as n increases. The result of flipping a coin, for example, is unpredictable, yet there is statistical regularity in the ratio m/n approaching 1/2 for either heads or tails (Figure 6.1).

Figure 6.1. The 'mathematics' gallery of BITM (Inaugurated on 8 May, 2010).

Source: Image by Wikimedia commons.

Noise in radio signals, the intensity of wind gusts, mechanical vibration due to atmospheric disturbances, Brownian motion of particles in a liquid, the number of telephone calls made by a given population, the length of queues at a ticket counter, the choice of transportation modes by a group of individuals, and countless other random phenomena abound in scientific fields.

Any credible conceptual model of a real-world occurrence contains unpredictability. Statistical concepts and methods are not only valuable, but they are frequently required to comprehend the world around us. They provide new insights into the behavior of many phenomena that you may meet in your chosen engineering or science field of specialty.

In the face of uncertainty and variance, statistics teaches us how to make intelligent judgments and informed decisions. There would be no need for statistical procedures or statisticians if there was no uncertainty or variance.

A single observation would reveal all desired information if every component of a particular type had the same lifetime, all resistors produced by a particular manufacturer had the same resistance value, pH determinations for soil specimens from a particular location gave identical results, and so on.

6.2. ORGANIZATION OF TEXT

This book focuses on the creation of fundamental ideas for building probability models and their subsequent examination. The basic cycle of this endeavor, like other scientific modeling techniques, consists of a number of fundamental steps.

A basic understanding of probability theory and random variables is essential to the modeling process because they provide the necessary mathematical machinery for carrying out the procedure and deducing the results. The induction process is when the model's structure is derived from genuine observations of the scientific topic under investigation (Figure 6.2).

Figure 6.2. BITM mathematics gallery.

Source: Image by Wikimedia commons.

Statistical inference encompasses model verification and parameter estimate using observed data (E). Insufficient inductive reasoning or insufficient or deficient data may cause a model to be rejected at this point. It's possible that factual observations or further data will need to be reexamined in this case.

Finally, after model substantiation, model analysis and deduction are performed to obtain the desired results. The book is organized into two sections in accordance with this fundamental step plan.

6.3. PROBABILITY TABLES AND COMPUTER SOFTWARE

Calculations of various probabilities and statistical functions will be required to apply the materials in this book to actual applications, which might be time consuming. Some probability tables are included in Appendix A to help with these computations (Figure 6.3).

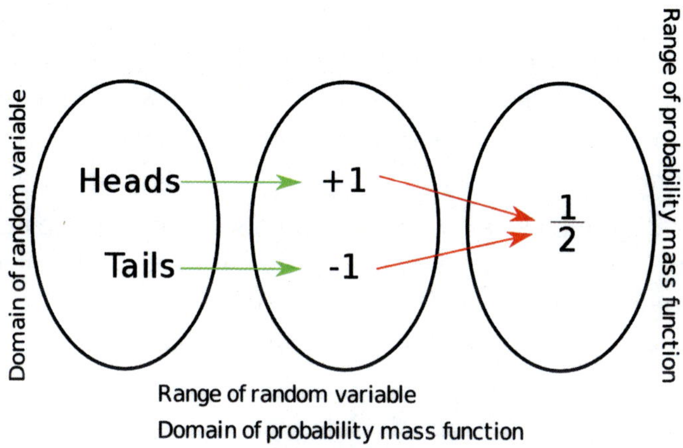

Figure 6.3. Illustrates the concept of random variable.

Source: Image by Wikimedia commons.

However, a great range of computer software programs and spreadsheets that supply this information as well as perform a variety of other statistical calculations are now accessible. Appendix B, for example, lists several statistical functions accessible in Microsoft Excel™ 2000.

6.4. PROBABILITY AND RANDOM VARIABLES

The essential tools for developing and accessing mathematical models for random processes are provided by the mathematical theory of probability. When random phenomena is researched, an experiment is conducted whose outcome cannot be predicted in advance. Experiments of this type spring to mind instantly when thinking of games of chance.

In fact, questions like these stimulated the early development of probability theory in the fifteenth and sixteenth centuries (Todhunter, 1949). Random phenomena are used to explain a wide range of circumstances in science and engineering. They can be divided into two categories in general.

The first category is concerned with unclear physical or natural events. Complexity, our lack of understanding of all the causes and effects, and a lack of information all contribute to uncertainty in issue formulation. Take, for example, weather forecasting. Satellite tracking data and other meteorological data are simply insufficient to make a credible forecast of what weather conditions will prevail in the coming days.

It's easy to see why weather reports on the radio and television (TV) are given in probabilistic terms. The second type of problem that has been extensively studied using probabilistic models is one which involves variability. Consider a traffic flow situation in which an engineer wants to know the number of vehicles crossing a specific spot on a road in a given amount of time.

This quantity fluctuates randomly from one interval to the next, and this variation represents unpredictable driver behavior and is a problem in and of itself. This trait pushes us to take a probabilistic approach, and probability theory is a useful tool for studying such problems (Figure 6.4).

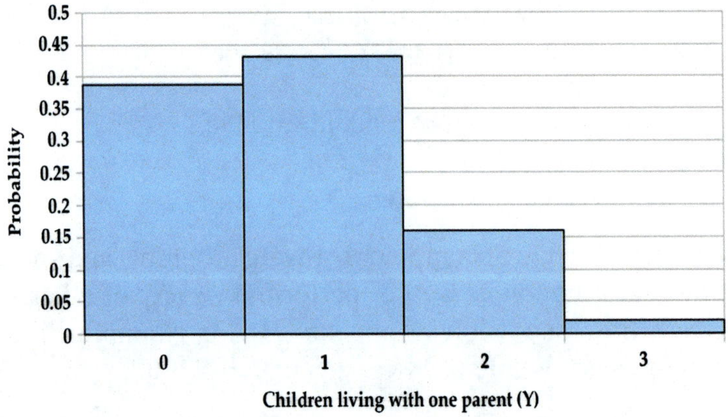

Figure 6.4. Probability histogram for random variable.

Source: Image by Wikimedia commons.

It is reasonable to model all real events with uncertainty and unpredictability, so it is only natural that probabilistic modeling and analysis play a major role in the study of a wide range of topics in science and engineering. There is no question that the use of probabilistic formulations will become more common in most scientific areas in the future.

6.4.1. Sample Space and Probability Measure

In probability theory, consider a "random experiment," which is an experiment where the results depend on chance. It is assumcd that all of the different possible outcomes of a random experiment are known and are part of a basic set called the sample space. Each possible outcome is called a "sample point," and an event is usually defined as a subset of the sample space that contains one or more sample points.

It's important to note that the sample space for a random experiment is not always the same. It depends on the point of view and the questions that need to be answered. For example, a business that makes industrial goods is making 100 resistors. Because of the way they are made and how they are measured, their values can be anywhere from 99 to 101.

A measurement of a resistor is a random experiment, and the possible results can be described in different ways depending on why the experiment is being done. If something is acceptable and something else is not, it is enough to say that the sample space has two elements: "acceptable" and "unacceptable." From the point of view of a different user, however, possible, 99.5–100, 100–100.5, and 100.5–101.

In this case, there are four sample points in the sample space. Lastly, if each possible reading is a possible outcome, the sample space is now a real line from 99 to 101 on the ohm scale. There are an uncountably infinite number of sample points, and the sample space is a set with no numbers. Consider a negotiation between the United States and another country about energy.

This will show that a sample space is not fixed by the act of doing the experiment, but by the point of view of the observer. From the US government's point of view, the only possible outcomes may be success or failure.

For the consumer, though, a more direct set of possible outcomes may include price increases and decreases for gasoline. The way sample space, sample points, and events are described shows that they fit well into the framework of set theory. Set theory is a framework for analysing the results of a random experiment.

6.4.2. Assignment of Probability

The axioms of probability tell us what a probability measure is and how it works, which is in line with what is known from experience. But they don't tell us how likely different things are to happen. In applied sciences, looking at how often something happens is a natural way to figure out how likely it is to happen.

If a random experiment is done a lot of times, let's say n, then for any event A, let nA be the number of times A happened in the n trials and define the ratio nA/n as how often A happened. As n gets bigger, it is likely that this ratio will move toward a unique limit when things are stable or follow a statistical pattern.

This limiting value of the relative frequency is a good candidate for the probability of A because it has all the properties that a probability measure needs. This way of thinking is used, for example, to say that the chance of getting "heads" when you flip a coin is 1/2.

The relative frequency method is often used in science and engineering because it is based on facts and is consistent with the axioms from Section 2.2.1. Relative likelihood is a common but more subjective way to figure out how likely something is. If you can't do an experiment a lot of times or can't do it at all, you may have to make a subjective decision about how likely something is to happen.

In this way of looking at things, the statement "There is a 40% chance that it will rain tomorrow" means that the number 0.4 is based on the information available and professional judgment. In most of the problems in this book, one of the two methods is used to figure out the odds of some simple but basic things happening. The theory of probability is then used to figure out other interesting probabilities.

6.4.3. Conditional Probability

Probabilities are given to different events based on what is known about the experiment at the time the probabilities are given. After the first assignment, some information about how the experiment will turn out may become available.

6.4.4. Probability Plots

An investigator will often have a sample of numbers and want to know if it is likely that they came from a certain type of population distribution (e.g., from a normal distribution). For one thing, many formal statistical inference procedures are based on the idea that the population is spread out in a certain way.

This kind of procedure shouldn't be used if the real probability distribution is very different from the one that is assumed. For example, the article "Toothpaste Detergents: A Potential Source of Oral Soft Tissue Damage" (Intl. J. of Dental Hygiene, 2008: 193–198) says, "The data were assumed to be normally distributed because the sample size for each experiment (replication) was limited to three wells per treatment type."

As a reason for this leap of faith, the authors wrote, "Descriptive statistics showed standard deviations that suggested a normal distribution was likely." This argument doesn't make a lot of sense. Also, knowing the underlying distribution can sometimes give you a better idea of how the data were made physically.

Putting together a "probability plot" is a good way to check a distributional assumption. If the distribution on which the plot is based is correct, the points on the plot should be close to a straight line. If the real distribution is very different from the one that was used to make the plot, the points will probably not follow a straight line.

6.5. RANDOM VARIABLES AND PROBABILITY DISTRIBUTIONS

So, when the experiment is done, each result is known by the real number it was given instead of its physical description. For example, when the only possible results of a random experiment are success and failure, it is given that the event "success" the number one and the event "failure" the number zero. Instead of success and failure, the sample points in the associated space are now 1 and 0. This means that when you say "the outcome is 1," you mean "the outcome is success." This procedure not only lets us change a sample space with arbitrary elements into a new sample space with only real numbers, but it also lets us use arithmetic to figure out how likely something is to happen.

Also, most problems in science and engineering are about how to measure something. Because of this, the sample spaces for many interesting random experiments are already sets of real numbers. So, the way real numbers are assigned is a natural way to bring everything together.

On the basis of this, a variable can be used to represent real numbers whose values depend on the results of a random experiment. This gives rise to the idea of a random variable, which will be explained in more detail below.

6.5.1. Random Variables

Think about a random experiment where the results are parts of the sample space in the probability space that underlies it. To make a model for a random variable, it is assumed that each outcome can be given a real number by following a certain set of rules.

It can be seen that the "number" is really a point function with real values that is defined over the basic probability space. If the random variable is defined over a sample space with a finite or countably infinite number of sample points, it is called a continuous random variable.

In this case, the random variable has discrete values, and it is possible to list all of those values. When a sample space has an uncountably infinite number of sample points, the associated random variable is called a random variable, and its values are spread out over one or more continuous intervals on the real line.

This distinction is made because there are different things to think about when assigning probabilities to each. Both kinds of random variables are important in science and engineering.

6.5.2. Probability Distributions

The way the probabilities are spread out over the values a random variable takes on describes how it acts. For a discrete random variable, you can describe this distribution in two ways: with a probability distribution function or a probability mass function.

They are the same in the sense that knowing either one tells you everything you need to know about the random variable. For a continuous random variable, the functions that are the same as for a discrete random variable are the probability distribution function and the probability density function.

6.5.3. Two or More Random Variables

In many situations, it makes more sense to use two or more numbers at once to describe the result of a random experiment. For example, describing the weight and height of a given population, studying the changes in temperature and pressure during a physical experiment, or figuring out how the temperatures in a given area change over the course of a year.

In these situations, consider two or more random variables together and try to describe how they behave together. First, let's look at the case where there are two random variables, X, and Y. Their joint probability distributions have been defined in the same way as for a single random variable.

It can be seen that the random variables X and Y can also be thought of as parts of a random vector with two dimensions, say Z. Bivariate distributions are sometimes used to talk about the joint probability distributions of two random variables. As we'll see, it's easy to apply this to situations with more than two random variables, which are called "multivariate distributions."

6.5.4. Functions of Random Variables

The main topic of this chapter is figuring out how the probability distributions of two random variables, X, and Y, relate to each other when they are linked by Y g. (X). The functional form of g(X) is given and can be calculated with certainty. In the case of many random variables, the joint probability distribution of Y j, j 1, 2...., m, depends on Xk, k 1, 2...., n.

Transformations of random variables are a type of problem that has been talked about in Chapter 4 in more than one place. For example, the example looks at the transformation Y X1 Xn, and another example looks at the transformation of 3n random variables (X1, X2...., X3n) to two random variables (X0, Y0) defined by some equations.

Most phenomena in science and engineering are based on functional relationships, in which one or more dependent variables are expressed in terms of one or more independent variables. For example, force depends on cross-sectional area and stress, distance traveled in a certain amount of time depends on speed, and so on.

So, the techniques are used to figure out the probabilistic behavior of random variables that depend on other random variables whose probabilistic properties are already known. In what comes next, random variables will be changed in a systematic way.

6.6. SOME IMPORTANT DISCRETE DISTRIBUTIONS

This talks about some distributions of discrete random variables that are important as models of scientific phenomena. Considering what these distributions are and how they can be used. When it is known that how these random variables show up, the right distribution can be chosen for the scientific that is being considered. In this step, a model is chosen based on what is known about the physical phenomenon being studied.

6.6.1. Bernoulli Trials

Repeating a random experiment with the same basic rules can be used to describe a wide range of real-world situations: a series of trials are done so that (a) for each trial, there are only two possible outcomes, such as success or failure; (b) the odds of these outcomes happening stay the same throughout the trials; and (c) each trial is done on its own.

Bernoulli trials are tests that are done under these conditions. Even though the situation is simple, the mathematical models that come from this basic random experiment can be used in many different ways.

6.6.2. Binomial Distribution

A lot of people are interested in the probability distribution of a random variable X that represents the number of successes in a series of n Bernoulli trials, no matter what order they happen in. It is clear that X is a discrete random variable, since it can take on any of the values 0, 1, 2…., n. To figure out its probability mass function, look at pX(k), which is the chance of getting exactly k successes out of n tries.

This can happen in as many ways as there are k letters S and n boxes. Now, there is n possible positions for the first S, n 1 possible positions for the second S, etc., and n k 1 possible positions for the kth S. So, there are n(n – 1)…. (n – k + 1) possible ways to set things up.

6.6.3. Poisson Distribution

In this, a distribution has been described that can be used in a wide range of physical settings. It is used in mathematical models to describe things that happen in a certain amount of time, like the release of particles from a radioactive substance, the arrival of passengers at an airport, the spread of dust particles in a certain space, the arrival of cars at an intersection, and many other similar things.

Based on this result, there can be an important point. It means that if a random variable X is Poisson distributed with parameter, then a random variable Y is also Poisson distributed with parameter p if it is made by picking only one of the items counted by X with probability p.

This result can also be used when Y is the number of disaster-level hurricanes and X is the total number of hurricanes that happen in a given year, or when Y is the number of people who can't get on a flight because it's already full and X is the number of passengers who arrive.

6.6.4. Spatial Distributions

The Poisson distribution is based on how arrivals change over time, but the same argument works for how points are spread out in space. Think about how the flaws in a material are spread out.

The number of flaws in a given volume has a Poisson distribution if Assumptions 1–3 are true, replacing time intervals with volumes, and if it is reasonable to assume that the probability of finding k flaws in any region depends only on the volume and not on the shape of the region.

The Poisson distribution is also used to figure out how many bacteria are on a Petri plate, how fertilizers are spread in a field by an airplane, and how industrial pollutants are spread in a certain area.

6.7. SOME IMPORTANT CONTINUOUS DISTRIBUTIONS

Let's look at some important continuous probability distributions. Continuous random variables are used to model physical quantities like time, length, area, temperature, pressure, load, intensity, etc., when they need to be described in terms of chance.

6.7.1. Uniform Distribution

The uniform distribution is one of the simplest, and it is often used when there is no reason to give different probabilities to different ranges that a random variable could take over a given interval.

For example, the arrival time of a flight might be considered uniformly distributed over a certain amount of time, or the distance from the location of live loads on a bridge to an end support might be well represented by a uniform distribution over the bridge span (Figure 6.5).

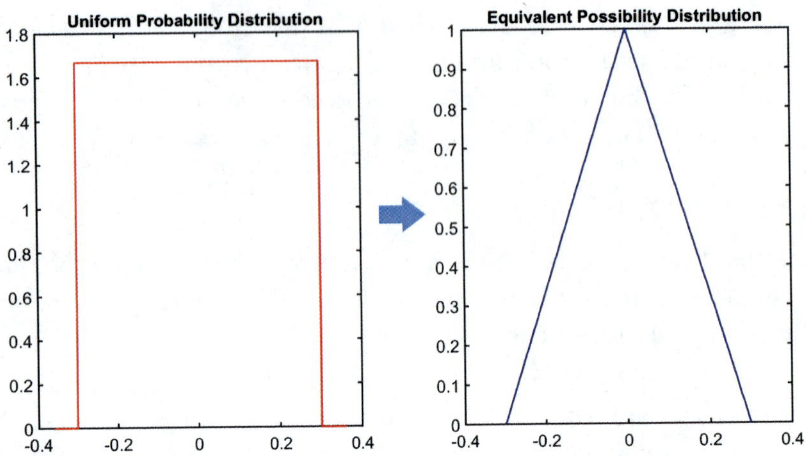

Figure 6.5. Uniform distribution in probability and possibility.

Source: Image by Wikimedia commons.

Let's also say that people often give a random variable a uniform distribution just because they don't know much about it beyond the range of values it has.

6.7.2. The Normal Distribution

The most important distribution in probability and statistics is the normal distribution. A normal curve can be used to fit the distributions of a lot of groups of numbers very well.

Examples include heights, weights, and other physical traits (the famous 1903 Biometrika article "On the Laws of Inheritance in Man" talked about many of these), measurement errors in scientific experiments, anthropometric measurements on fossils, reaction times in psychological experiments, measurements of intelligence and aptitude, scores on different tests, and many economic measures and indicators (Figure 6.6).

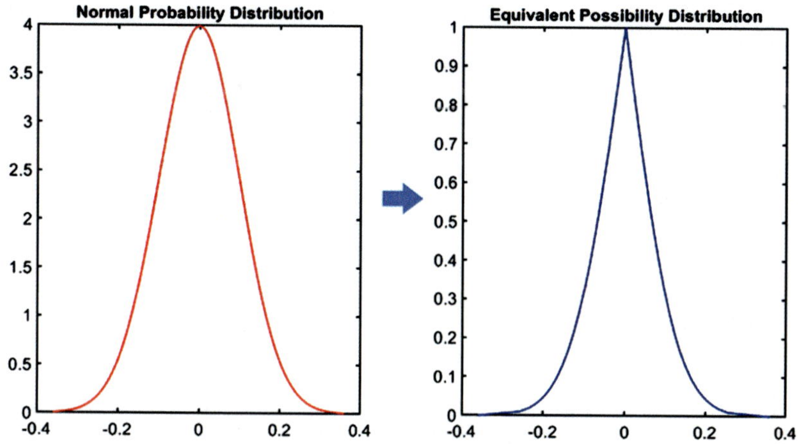

Figure 6.6. A normal probability distribution and its equivalent distribution in possibility domain done using probability possibility transformation.

Source: Image by Wikimedia commons.

6.7.3. The Central Limit Theorem

The powerful central limit theorem shown below explains why the normal distribution is so useful in real life. Instead of giving the theorem in its most general form, it works well for us to give a more specific version that Lindberg came up with (1922). The central limit theorem talks about a very large group of random events for which the normal distribution is a good approximation.

In other words, the randomness of a physical phenomenon tends toward a normal distribution when it is made up of many small random effects that add up. This is true regardless of the distributions of the individual effects. For example, even though all cars of the same brand are supposed to be made in the same way, the amount of gas they use varies from one car to the next.

This randomness comes from a lot of different things, such as inaccuracies in the manufacturing process, differences in the materials used, differences in weight and other specs, differences in the quality of the gasoline, and different driving styles (Figure 6.7).

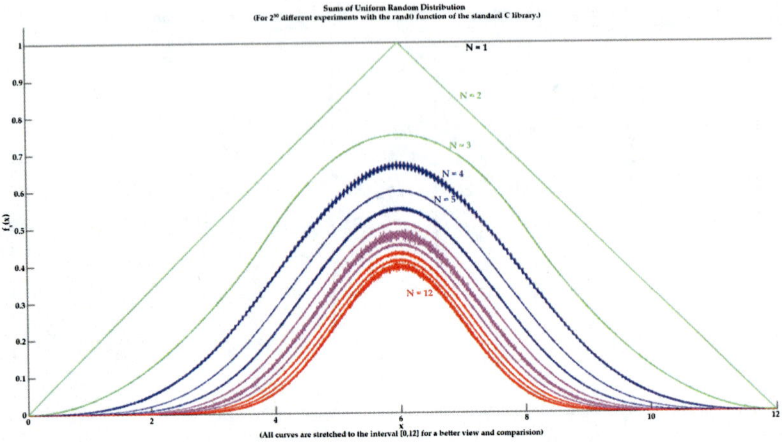

Figure 6.7. This file shows plotting of twelve different probability distribution functions formed by summing the standard C language library function rand(), for illustrating the Central Limit Theorem.

Source: Image by Wikimedia commons.

If it is assumed that each of these differences contributes to the randomness of gasoline consumption, the central limit theorem tells us that it tends to follow a normal distribution. Normal distributions can also be used to get a good idea of the temperature changes in a room, the readout errors of an instrument, the target errors of a certain weapon, and so on.

6.7.4. Lognormal Distribution

It has been seen that normal distributions are made up of many random events added together. Now, think about another common thing that happens because of many random effects that add up. One example of a multiplicative effect is how internal damage to a material at a certain stage of loading is a random percentage of damage at the previous stage.

In biology, the way an organism's size changes over time is another example of how growth is affected by many small changes that are proportional to the size at the time. Other examples include the size distribution of particles under impact or impulsive forces, the life distribution of mechanical parts, the distribution of personal incomes due to annual adjustments, and other similar things.

6.8. OBSERVED DATA AND GRAPHICAL REPRESENTATION

Let's consider this problem of estimating parameters with the data that is provided. In this discussion, let's discuss about how to figure out if a model is reasonable and how to choose a model from a number of competing distributions when none of them is better based on the physical properties of a certain phenomenon.

Let's make it clear right away that the problem of estimating parameters is just that: an estimation problem. This is because the situation is based on probabilities. A group of observations, say n of them, is a sample of the values of the random variable that they represent. If the sequence of n observations is repeated, the randomness of the experiment should lead to a different set of values that were observed.

So, any reasonable rule for getting parameter estimates from a set of n observations will give different estimates for different sets of observations. In other words, you can't expect true parameter values to come from a single set of observations with a limited number of points.

The main thing that must be done, then, is to get useful information about the distribution parameters by observing the random process at work and using the numbers in a systematic way.

6.8.1. Histogram and Frequency Diagrams

Given a set of independent observations x 1, x 2...., and x n of a random variable X, it's a good idea to organize and present them in a way that makes them easy to understand and evaluate. When there are a lot of observed data, a histogram is a great way to show the data graphically.

It helps with (a) figuring out if the assumed model is good enough, (b) figuring out what the percentiles of the distribution are, and (c) figuring out what the distribution parameters are. Let's look at an example of a chemical process that makes batches of a wanted material (Hill, 1975). The values of the samples range from 64 to 76 (Figure 6.8).

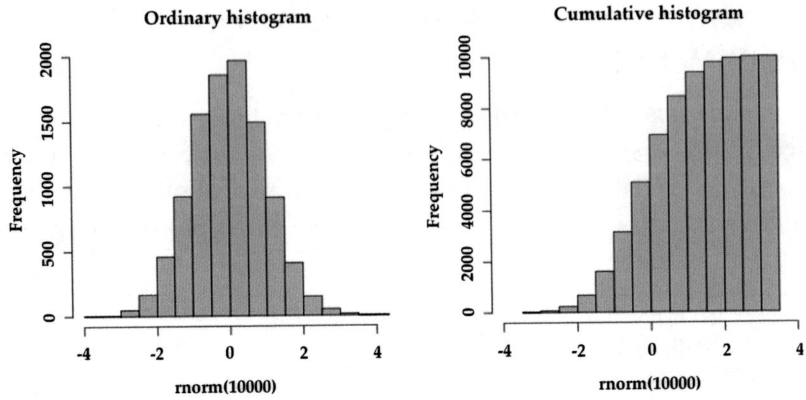

Figure 6.8. A plot showing a regular and a cumulative histogram of the same data.

Source: Image by Wikimedia commons.

The histogram can be made by dividing this range into 12 equal intervals and plotting the total number of observed yields in each interval as the height of a rectangle over the interval. When the histogram's ordinate is divided by the total number of observations, which in this case is 200, and by the interval width, you get a frequency diagram (which happens to be one in this example).

It can be considered that the histogram or frequency diagram gives us a quick idea of the observed data's range, relative frequency, and scatter. In the case of a discrete random variable, the histogram and frequency diagram are in the form of a bar chart, while in the case of a continuous random variable, they are in the form of connected rectangles.

Take, for example, the number of accidents each driver in California had over the course of six years. It's important to say something about how the number of intervals was chosen for the histograms and frequency diagrams.

For this example, the choice of 12 intervals makes sense because the observations cover a wide range of values and the resulting resolution is good enough for the probability calculations that have already been done.

For the same example, a histogram is made with 4 intervals instead of 12. It's clear that it gives a very different and less accurate picture of how data behaves. So, it's important to pick the right number of intervals based on what you want to learn from the mathematical model.

6.9. INTRODUCTION TO STATISTICS

In this scenario, it is understood that to learn about something, you must first gather information. Statistics is the study of how to learn from numbers. It is about gathering information, describing it, and figuring out what it means, which often leads to drawing conclusions.

6.9.1. Data Collection and Descriptive Statistics

Sometimes, a statistical analysis starts with a set of data that is already known: For example, the government collects and makes public information about how much rain falls each year, how often earthquakes happen, the unemployment rate, the gross domestic product, and the rate of inflation. Statistics can be used to describe, sum up, and look at these pieces of information.

In other cases, there are no data yet, so statistical theory can be used to plan an experiment that will generate data. How the data will be used should determine which experiment is chosen. For example, say a teacher wants to know which of two ways to teach computer programming to beginners is the best.

To study this question, the teacher might split the class into two groups and teach each group in a different way. At the end of class, the students can take a test, and the scores from each group can be compared. If the data, which are the test scores of the people in each group, show that one group did much better, it might seem reasonable to think that the way that group was taught is better (Figure 6.9).

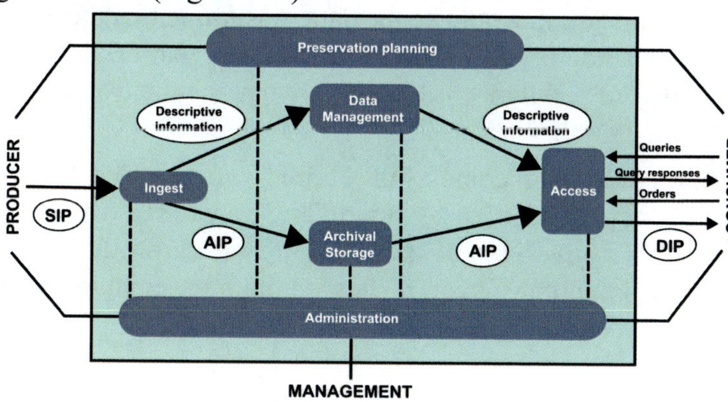

Figure 6.9. OAIS functional model.
Source: Image by Wikimedia commons.

It is important to note, though, that in order to draw a valid conclusion from the data, the students had to be put into groups so that neither group was more likely to have the students who were better at programming by nature. For example, the teacher shouldn't have split the class into two groups, one for men and one for women.

If that were the case, even if the women did much better than the men, it wouldn't be clear if it was because of the way they were taught or because women may be better at learning programming skills in general. Most people agree that the best way to avoid this mistake is to split the class into two groups "at random." This means that the people in a group are split up in such a way that any choice they could make is just as likely as any other.

At the end of the experiment, you should talk about what you found. For example, the two groups' scores should be shown. Also, summary measures, like the average score of each group's members, should be shown. Descriptive statistics is the part of statistics that deals with describing and summarizing data.

6.9.2. Inferential Statistics and Probability Models

A probability model for the data is the set of all these assumptions taken together. The form of the probability model that is assumed can sometimes be guessed from the nature of the data. For example, suppose an engineer wants to know how many of the new computer chips made with a new method will be broken.

The engineer might choose a group of these chips, and the number of broken chips in this group would be the data. Assuming that the chips chosen were chosen "at random," it is reasonable to think that each one is broken with a probability of p, where p is the unknown percentage of all chips made with the new method that are broken.

The information that comes out of this can then be used to draw conclusions about p. In other cases, it won't be easy to see which probability model is best for a given set of data. But if it is carefully described and present the data, sometimes it can be figured out a reasonable model, which can then be tried to confirm with more data (Figure 6.10).

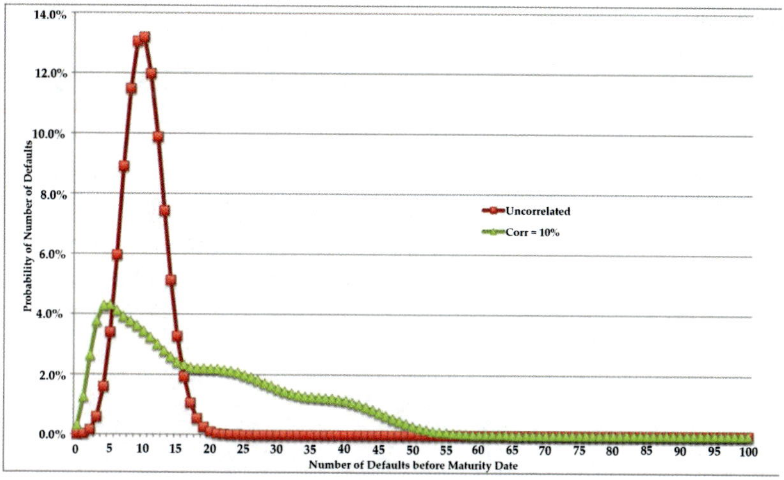

Figure 6.10. Probability distribution around mean default probability of 10%, N=100, rho of 0% and 10%. Created using Gaussian Copula model and 5,000 simulations.

Source: Image by Wikimedia commons.

Because statistical inference is based on making a probability model to describe the data, you need to know something about the theory of probability to understand statistical inference. In other words, statistical inference starts with the idea that important parts of the thing being studied can be described in terms of probabilities. It then uses data to make inferences about these probabilities and come to conclusions.

6.9.3. Populations and Samples

Most of the time, the population is too big for us to look at each person. For example, there are people who live in a certain state, or all the TVs made in the last year by a certain company, or all the homes in a certain community. In these situations, try to learn about the whole population by picking and studying a small part of it. A sample is a small part of a whole population.

If the sample is to tell us something about the whole population, it must be, in some way, like the population as a whole. For example, let's say it is required to find out how old the people who live in a certain city are, and the ages of the first 100 people is provided who go to the town library.

If the average age of these 100 people is 46, 2 years, is it considerable that this is about the average age of the whole population? Most likely not, because the arguments can be placed as that the sample chosen in this case is probably not representative of the whole population. This is because the library is usually used more by students and older people than by people of working age.

In some cases, like the library example, a sample is provided and then it is required to decide if this sample is a good representation of the whole population. In real life, it can't be usually assumed that a given sample is representative of a population unless that sample was chosen at random (Figure 6.11).

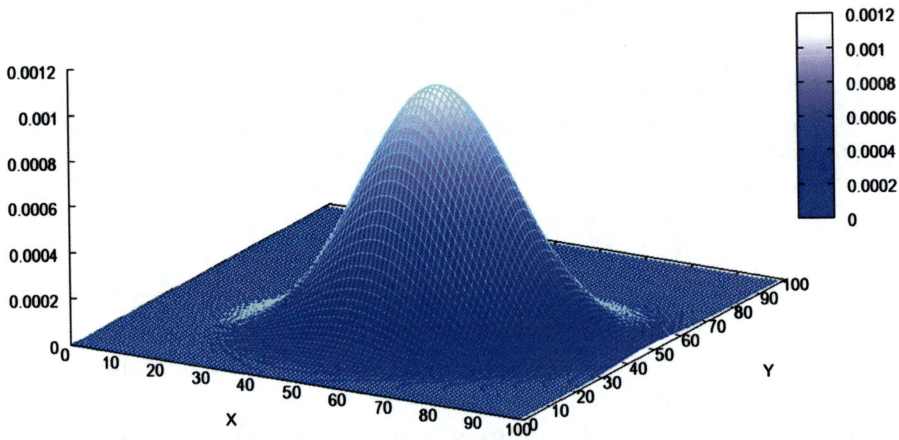

Figure 6.11. Multivariate gaussian.

Source: Image by Wikimedia commons.

This is because any non-random rule for choosing a sample often leads to a sample that is biased toward some data values over others. So, even though it may seem strange, the best way to get a representative sample is to choose its members at random, without thinking about the things that will be chosen in advance.

In other words, it is not required to try to pick the sample so that, for example, it has the same percentage of men and women and the same percentage of people in each job as the whole population. Instead, it must be left it to "chance" to get numbers that are close to right. Once a random sample is provided, statistical inference can be used to figure out what the whole population is like by looking at the sample.

6.9.4. A Brief History of Statistics

During the Renaissance, the Italian city-states of Venice and Florence started to keep track of information about their people and their economies in a systematic way. The word "statistics" came from the word "state," and it was used to mean a group of facts that were important to the state. From Italy, the idea of collecting data spread to the rest of Western Europe.

In fact, by the first half of the 1600s, it was common for governments in Europe to demand that parishes record births, marriages, and deaths. This last statistic was especially interesting because of the bad state of public health. Before the 19th century, Europe had a very high death rate. This was mostly because of epidemic diseases, wars, and famines.

The worst epidemics were the plagues. Beginning with the Black Plague in 1348, there were many plagues for almost 400 years. In 1562, the City of London started putting out weekly bills of deaths as a way to get the King's court to think about moving to the country. At first, these bills of death listed where people died and if they died from the plague.

Beginning in 1625, all causes of death were added to the bills. John Graunt, an English merchant, wrote a book called Natural and Political Observations Made upon the Bills of Mortality. It came out in 1662.

During the 19th century, mathematicians like Jacob Bernoulli, Karl Friedrich Gauss, and Pierre-Simon Laplace worked on probability theory. However, it was almost never used to study statistical findings, because most social statisticians at the time were happy to let the data speak for themselves.

At that time, statisticians were not interested in making conclusions about specific people. Instead, they were interested in society as a whole. So, they didn't care about samples; instead, they tried to get censuses of the whole population. Because of this, almost no one in the 19th century used probabilistic inference from samples to the whole population.

Statistics didn't become concerned with drawing conclusions from numbers until the late 1800s. Francis Galton's work on analyzing hereditary genius with what is referred to as regression and correlation analysis started the movement, and Karl Pearson's work gave it a lot of momentum (Figure 6.12).

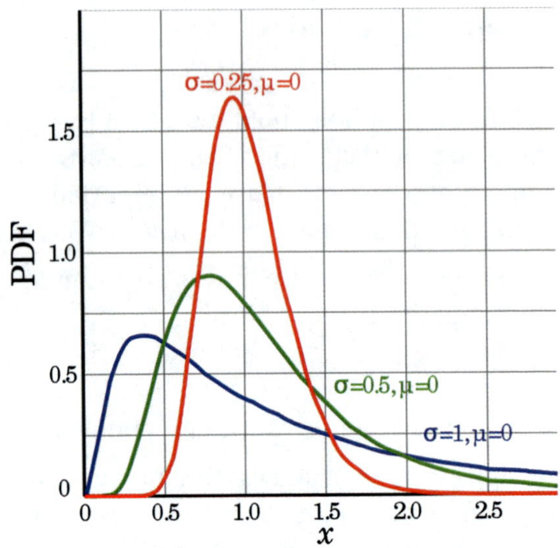

Figure 6.12. Probability distribution functions of log-normal distributions.

Source: Image by Wikimedia commons.

Pearson made the chi-square goodness of fit tests and was the first director of the Galton Laboratory, which was given to him by Francis Galton in 1904. There, Pearson started a research program to find new ways to use statistics to draw conclusions. His lab was open to advanced students in science and business who wanted to learn statistical methods they could use in their own fields.

W. S. Gosset, a chemist by training, was one of the first people to work with Pearson. He showed how much he cared for Pearson by publishing his own works under the name "Student." (According to a well-known story, Gosset was afraid to publish under his own name because he didn't want his bosses at the Guinness brewery to be upset that one of their chemists was doing research in statistics.) Gosset is known for coming up with the t-test.

At the beginning of the 20th century, population biology and agriculture were two of the most important fields where statistics were used. This happened because Pearson and others in his lab were interested, and also because of what the English scientist Ronald A. Fisher had done.

The theory of inference that these pioneers, like Karl Pearson's son Egon and the Polish-born mathematician and statistician Jerzy Neyman, came up with could be used to solve a wide range of numerical and practical problems.

Because of this, after the first few decades of the 20th century, a growing number of people in science, business, and government began to see statistics as a tool that could help them solve scientific and practical problems in a quantitative way. Statistics are talked about everywhere these days.

Every newspaper and magazine have statistics that describe things. Statistical inference is now needed in public health and medical research, engineering, and scientific studies, marketing, and quality control, education, accounting, economics, weather forecasting, polling, and surveys, sports, insurance, gambling, and any research that claims to be scientific. Statistics has become an important part of our intellectual history.

6.10. DESCRIPTIVE STATISTICS

Descriptive statistics are used to describe the basic features of the data in a study. They provide simple summaries about the sample and the measures. Together with simple graphics analysis, they form the basis of virtually every quantitative analysis of data.

Descriptive statistics are typically distinguished from inferential statistics. With descriptive statistics you are simply describing what is or what the data shows. With inferential statistics, you are trying to reach conclusions that extend beyond the immediate data alone.

6.10.1. Branches of Statistics

After collecting data, a researcher may just want to sum up and describe the most important parts. To do this, you need to use techniques from descriptive statistics. Some of these methods involve making graphs. Histograms, boxplots, and scatter plots are good examples.

Other ways to describe something involve figuring out numbers like means, standard deviations, and correlation coefficients to give an overall picture. Now that statistical software is widely available, it is much easier to do these things than it used to be.

When it comes to doing math and making pictures, computers are much better than people, as long as the user gives them the right instructions. This means the investigator won't have to spend as much time on "grunt work" and will have more time to look at the data and figure out what's important.

6.10.2. The Scope of Modern Statistics

Statistical methods are used by researchers in almost every field today, such as

- molecular biology (analysis of microarray data);
- Statistical methods are used by researchers in almost every field today, such as molecular biology (analysis of microarray data);
- materials engineering (studying properties of various treatments to retard corrosion);
- marketing (developing market surveys and strategies for marketing new products);
- health care (identifying sources of diseases and ways to treat them); and
- civil engineering (assessing the effects of stress on structural elements and the impacts of traffic flows on communities).

6.11. ENUMERATIVE VERSUS ANALYTIC STUDIES

In the 1950s and 1960s, W. E. Deming, a statistician who played an important role in Japan's quality revolution, distinguished between enumerative and analytic studies. A population is the collection of individuals or objects that are a finite, identifiable, unchanging set of individuals or objects that makes up a population, where the focus is on the finite, identifiable, and unchanging individual or object that makes up the population.

It is either possible for an investigator to construct a sampling frame, which is a list of individuals or objects that will be sampled. A frame might be used, for example, to determine whether the number of valid signatures exceeds a specified value on a petition to qualify a political initiative for the ballot in a upcoming election; a sample, typically, would be used to determine whether the number of valid signatures exceeds a specified value.

For instance, the frame could contain the serial numbers of all furnaces produced by a specific company during a specific period of time; it may be possible to select a sample of these furnaces so that something about their average lifetime can be determined.

Generally, the use of inferential methods to be developed in this book is not controversial in such settings (although statisticians may disagree over which particular methods should be used). An analytical study is generally defined as one that is not enumerative in nature.

It has been found that such studies are often conducted with the objective of improving a future product by taking an action on a process of some sort (e.g., recalibrating equipment, or adjusting the activity of a catalyst or some other input). The data can be obtained mainly from existing processes, which are often very different from those to be developed in the future.

Consequently, there is no sampling frame that identifies the individuals or objects that are of interest. For example, a sample of five turbines with a new design may be manufactured and tested experimentally to determine the efficiency of the turbine.

Several of these prototypes can be considered representative of a conceptual population of all prototypes that can potentially be manufactured under similar conditions, but they may not necessarily be a representative sample of the units produced once regular production begins. Using sample information in order to draw conclusions about future production units may present a problem.

It would be ideal to have an expert in the field of turbine design and engineering (or whatever other subject area may be relevant) assess the feasibility of such extrapolation. It is possible to get a good understanding of these issues by reading Gerald Hahn and William Meeker's article "Assumptions for Statistical Inference" (The American Statistician, 1993: 1–11).

6.12. COLLECTING DATA

Statistics deals not only with the arrangement and analysis of data once it has been acquired but also with the development of procedures for obtaining the data. If data is not adequately obtained, an investigator may not be able to answer the questions under discussion with a sufficient degree of confidence.

One recurrent concern is that the target population—the one about which conclusions are to be drawn—may be different from the population actually sampled. For example, advertisers would desire various kinds of information about the TV-viewing habits of potential customers.

The most systematic information of this sort comes from deploying monitoring devices in a small number of homes around the United States. It has been conjectured that placement of such devices in and of themselves modifies watching behavior, so that characteristics of the sample may be different from those of the target population.

When data collecting requires selecting individuals or items from a frame, the easiest technique for assuring a representative selection is to take a simple random sample. This is one for which any given subset of the stated size (e.g., a sample of size 100) has the same chance of being selected.

For example, if the frame consists of 1,000,000 serial numbers, the numbers 1, 2..., up to 1,000,000 might be placed on identical slips of paper. After placing these slips in a box and thoroughly mixing, slips might be extracted one by one until the appropriate sample size has been acquired.

Alternatively (and much to be preferred), a table of random numbers or a computer's random number generator could be deployed. Sometimes various sampling methods can be employed to make the selection process easier, to collect supplementary information, or to raise the degree of confidence in results.

One such method, stratified sampling, requires splitting the population units into nonoverlapping groups and obtaining a sample from each one. For example, a maker of digital versatile disc (DVD) players could require information about customer satisfaction for units produced during the preceding year. If three distinct models were created and sold, a separate sample may be selected from each of the three relevant strata.

This would result in information on all three models and verify that no one model was over-or underrepresented in the total sample. Frequently a "convenience" sample is generated by picking individuals or items without thorough randomization. As an example, a collection of bricks may be piled in such a way that it is extremely difficult for those in the middle to be selected.

If the bricks on the top and sides of the stack were somehow different from the others, ensuing sample data would not be representative of the population. Often an investigator will assume that such a convenience sample approximates a random sample, in which case a statistician's toolbox of inferential procedures can be applied; nonetheless, this is a judgment call.

6.13. CONCLUSION

In the conclusion of this chapter, it discusses about the significance of math, probability, and statistics in the field of engineering. It also discussed about the various probability tables and computer software that are used in the engineering. In this chapter, a couple of important discrete distributions such as Bernoulli trials, Bionomial distribution, Poisson distribution, and spatial distribution have also been discussed.

Towards the end of the chapter, it introduces to the concept of statistics, data collection and descriptive statistics, significance of populations and sample, branches of statistics, and the scope of modern statistics. In this chapter, enumerative versus analytic studies have also been discussed.

REFERENCES

1. Briš, R., (2022). *Probability and Statistics for Engineers* (p. 4). [e-Book] Available at: https://homel.vsb.cz/~bri10/Teaching/Prob%20&%20Stat.pdf (accessed on 1 July 2022).
2. Devore, J., (2010). *Probability and Statistics for Engineering and the Sciences* (p. 4). [e-Book] Available at: https://faculty.ksu.edu.sa/sites/default/files/probability_and_statistics_for_engineering_and_the_sciences.pdf (accessed on 1 July 2022).
3. Ross, S., (2022). *Introduction to Probability and Statistics for Engineers and Scientists* (5th edn., p. 3). [e-Book] Los Angeles, USA. Available at: https://minerva.it.manchester.ac.uk/~saralees/statbook3.pdf (accessed on 1 July 2022).
4. Soong, T., (2004). *Fundamentals of Probability and Statistics for Engineers* (p. 3). [e-Book] New York, USA. Available at: https://www.vfu.bg/en/e-Learning/Math--Soong_Fundamentals_of_probability_and_statistics_for_engineers.pdf (accessed on 1 July 2022).

CHAPTER 7

APPLICATIONS OF ENGINEERING ACROSS VARIOUS FIELDS

CONTENTS

To develop, design, and analyze solutions engineering is a profession in which people apply scientific theory. Engineering consists of major basic branches which all have numerous subdisciplines in general. Civil, chemical, software, mechanical, electrical, and industrial are the major branches of engineering.

To solve problems engineering is the application of science and mathematics. For scientific discoveries engineers figure out how things work and find practical uses. For innovations that advance the human condition, scientists, and inventors often get the credit, but in making those innovations available to the world it is engineers who are instrumental.

7.1. INTRODUCTION

Much of the modern society depends on engineered artifacts to function, but many members of modern society are not aware of the engineering techniques and practices that have developed the technology and infrastructure. Engineers' designs and creations are iPods, cell phones, airplanes, bridges, buildings, vehicles, computers, etc.

To the creation or modification of components, systems, and processes (which are often referred to as a product or an artifact) for the benefit of society, engineering is the application of the principles of mathematics and science.

To create such artifacts which represent a balance between quality, performance, and cost, engineers use a series of logical steps (the engineering design process). To succeed in the study of those subjects for a professional career in engineering, this chapter explores and examines the role and connections of math and science to engineering and the need.

7.2. TYPES OF ENGINEERING

7.2.1. Civil Engineering

It is a study and construction of infrastructure, such as roads, bridges, utilities, and buildings. Sit specialism like structural engineering and mechanical engineering within civil engineering. The buildings we live in and work in, the transportation facilities we use, the water we drink, and the drainage and sewage systems that are necessary to our health and well-being are all the effects of civil engineering (Figure 7.1).

Civil engineers:

- Measure and map the earth's surface.
- Design and supervise the construction of bridges, tunnels, large buildings, dams, and coastal structures.
- Plan, layout, construct, and maintain railroads, highways, and airports.
- For the control and efficient flow of traffic devise systems.
- River navigation and flood control projects building and planning.
- For water supply and sewage and refuse disposal provide plants and systems.

Figure 7.1. Civil engineers doing construction work.

Source: Image by Max Pixel.

Building may be a primal urge. To make us feel good about what we have built, our constructions, while they may be simply for shelter or transportation, often include esthetic touches that are there. Thus, to support weight, bridges have geometrical designs intended, but they also have an artistic detailing, in which they were built and a "look" that defines the era.

To develop the appearance of the structure.in constructing buildings, highways, and bridges, civil engineers work with architects. A building that falls down, or cannot be maintained, also represents a failure, but one that the civil engineer could possibly have prevented, ugly buildings represent a failed communication between the two professionals.

Erecting skyscrapers or bridges civil engineering is much more. In the interactions among structures, the earth, and water, with applications ranging from highways to dams and water reservoirs, civil engineers are trained. By the manufacturers of those materials, deeply involved with specifying appropriate construction materials, many civil engineers and others are also employed.

Elaborate planning, civil engineers can be outstanding project managers involved while constructing a large building or public-works project. Sometimes oversee thousands of workers and develop advanced computerization and planning policies are all done by engineers.

For solving the problems of society, and its history is intricately linked to advances in the understanding of physics and mathematics throughout history, civil engineering is the application of physical and scientific principles. Several specialized sub-disciplines include in civil engineering which is a broad profession, its history is linked to knowledge of structures, materials science, geography, geology, soils, hydrology, environmental science, mechanics, project management.

Such as stonemasons and carpenters, rising to the role of master builder throughout ancient and medieval history most architectural design and construction was carried out by artisans. By advances, knowledge was retained in guilds and seldom supplanted. Structures, roads, and infrastructure that existed were repetitive, and increases in scale were incremental.

Civil engineering is the work of Archimedes in the 3rd century BC, one of the earliest examples of a scientific approach to physical and mathematical problems applicable to it, including Archimedes Principle, which underpins our understanding of buoyancy, and practical solutions such as Archimedes' screw. Brahmagupta, an Indian mathematician, used arithmetic in the 7th century AD, it is based on Hindu-Arabic numerals, for excavation (volume) computations.

Many civil engineers are involved with preserving, protecting, or restoring the environment, most significantly. By civil engineers (in these two areas, many of them are known as environmental engineers) most water treatment and water purification projects are designed and constructed.

To clean up toxic industrial or municipal wastes at abandoned dump sites a growing number of civil engineers are involved in billion-dollar projects.

In diverse projects as preserving wetlands or beaches, maintaining national forest parks, and restoring the land around mines, oil wells, or factories civil engineers are engaged (Figure 7.2).

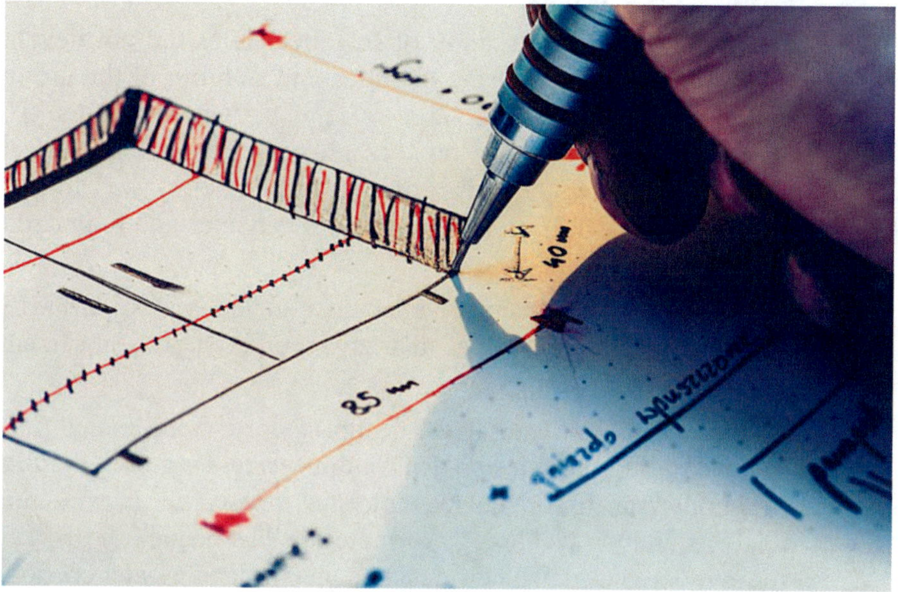

Figure 7.2. Civil engineering drawing.

Source: Image by Hippopx.

7.2.2. Application of Civil Engineering in Construction

This is divided into three categories: performance before construction (feasibility studies, site investigations, and design), performance during construction (dealing with clients, consulting engineers, and contractors), and performance after construction (maintenance and research).

- **Feasibility Studies:** Perhaps with alternatives, no major project today is started without an extensive study of the objective and without preliminary studies of possible plans leading to a recommended scheme. Feasibility studies may cover alternative methods—e.g., bridge versus tunnel, in the case of a water crossing—or, once the method is decided, the choice of route. Problems must be considered like both economic and engineering.
- **Site Investigations:** Once a plan has been adopted a more extensive investigation is usually imperative, when a preliminary

site investigation is part of the feasibility study. In constructional methods, money spent in a rigorous study of ground and substructure may save large sums later in remedial works or in changes made necessary.

It is surprising that a serious study of soil mechanics did not develop until the mid-1 since the load-bearing qualities and stability of the ground are such important factors in any large-scale construction, 1930s. The chief founder of the science is Karl von Terzaghi, gives the date of its birth as 1936, at Harvard University and an international society was formed, when the First International Conference on Soil Mechanics and Foundation Engineering was held.

In soil mechanics, today there are specialist societies and journals in many countries, and most universities that have a civil engineering faculty have courses.

- **Design:** The application of design theory from many fields required by the design of engineering—e.g., hydraulics, thermodynamics, or nuclear physics. Research in structural analysis and the technology of materials has opened the way for more rational designs, new design concepts, and greater economy of materials (Figure 7.3).

As more and more refined stress analysis of structures and systematic testing has been done, the theory of structures and the study of materials have advanced together. Structural designs can now be rigorously analyzed by computers, also modern designers not only have advanced theories and readily available design data.

Figure 7.3. Civil drawing and design.

Source: Image by piqsels.com.

- **Construction:** Most work is undertaken for large corporations, government authorities, and public boards and authorities but the promotion of civil engineering works may be initiated by a private client. For large specialized projects it is usual to employ consulting engineers many of these have their own engineering staffs.

To undertake feasibility studies, then to recommend a scheme and quote an approximate cost, the consulting engineer may be required first. The engineer is responsible, for the design of the works, supplying specifications, drawings, and legal documents in sufficient detail to seek competitive tender prices. Quotations and recommend acceptance of one of them is compared by the engineers.

The staff must supervise the construction and the engineer must certify completion of the work, although not a party to the contract, the engineer's duties are defined in it. The professional organizations exercise disciplinary control over professional conduct, actions must be consistent with duty to the client. The resident engineer is the consulting engineer's senior representative on the site.

The contractor undertakes to finance, design, specify, construct, and commission a project in its entirety is a phenomenon of recent years has been the turnkey or package contract. In this case, the consulting engineer is engaged by the contractor rather than by the client.

On the basis of the consulting engineer's specification and general drawings, the contractor is usually an incorporated company, which secures the contract. To any variations introduced and must approve the detailed drawings agreed by the consulting engineer.

- **Maintenance:** To the satisfaction of the consulting engineer, the contractor maintains the works. To ancillary and temporary works where these form part of the overall construction, responsibility for maintenance extended. By the contractor, a period of maintenance is undertaken, after construction and the payment of the final installment of the contract price is held back until released by the consulting engineer. Primarily with maintenance, central, and local government engineering and public works departments are concerned, for which they employ direct labor.
- **Research:** By government agencies, research in the civil engineering field is undertaken industrial foundations, the universities, and other institutions. The United States Bureau of

Standards and the National Physical Laboratory of Great Britain, there are countries which have government agencies involved in a broad spectrum of research, and establishments in building research, roads, and highways, hydraulic research, water pollution, and other areas. From research work promoted by industry many are government-aided but depend partly on income.

- **Environment Preservation:** On our planet, our actions have led to massive pollution. With the treatment of chemical, biological or thermal waste, and the purification of water and air dealt with environmental engineering is a sub-discipline of civil engineering. Environmental engineers administer pollution reduction, industrial ecology, and green engineering. The consequences of our actions on the is understandable by their help.
- **Urban Development:** For their municipal infrastructure, urban centers are known. Civil engineers are responsible for all of that from sidewalks to water supply networks, street lighting, sewer system, waste management systems. In the development of underground utility networks, the distribution of electrical and telecommunication networks, bus service networks, and more civil engineering plays a crucial role (Figure 7.4).

Figure 7.4. Urban development.

Source: Image by pixabay.com.

- **Water Resources Management:** The collection and management of water resources it is concerned with water resource engineering, a sub-discipline of civil engineering. this area of civil engineering finds useful applications.
- **Transportation Networks:** Sub-discipline of civil engineering that concerns itself with the efficient, conducive, and safe movement of goods and people from one point to another is known as transportation engineering. The design, construction, and maintenance of streets, rail systems, airports, mass transit, canals, and more are involved by it. We are thankful to have civil engineering.
- **Material Science:** Materials like ceramics, concrete, asphalt, strong metals like steel and aluminum, thermosetting polymers dependent by many of the world's civilization. To understand the fundamental properties of these important materials, material science is closely linked to civil engineering. Materials that will continue to have a greater impact in the future this sub-discipline has been at the forefront of nanoscience and nanotechnology.
- **Understanding Earthquakes:** From within the earth, the complex structures that cover our landscape have to be able to withstand forces., We can understand the interaction of structures to shaky grounds, thanks to earthquake engineering. To perform well in the event of an earthquake, this ensures that structures are built.
- **Surveying:** To help design a plan or map for construction, surveying is the process of analyzing and recording the characteristics of land areas span. Before construction commences, surveyors lay out the routes of highways, railways, pipelines, roads, streets, and the position of other infrastructures like harbor.

Roads, buildings, bridges, and water supply and sewage systems, civil engineers design and supervise construction of structures and infrastructure. For the elderly and physically challenged to all structures as well as infrastructure improvements for controlling and reducing urban environmental pollution of water and air, examples of new human related challenges are in providing ready access and easy mobility.

Del Webb Houses and the City of Phoenix, civil engineers typically work as consultants and for architectural and city organizations. In the design of roads and structures, they make use of mechanics from physics, but when addressing environmental issues related to water supply and sewage, also need the tools of chemistry and biology (Figure 7.5).

Figure 7.5. Surveying before construction.

Source: Image by Flickr.com.

We drive on to the buildings we reside in, the work of civil engineers surrounds us all, from the water we drink to the pavement. Within a community, complex transport systems connect points. Thanks to the hands of civil engineers' – road, air, sea, and rail networks span the entire globe. Trade, travel, the exchange of ideas and information, healthcare, education, industry, etc. are possible by civil engineering.

Our electricity demand has skyrocketed, over the past few decades and it will continue. Developing structures that consume as little energy as possible will be possible by civil engineers. Civil engineers will be responsible for building structures that will support us all, as the world population surges.

It's hard to imagine how the world we live in would look like without civil engineering, but we can be certain that marvelous man-made structures that fill our landscape will be absent. Civil engineering will continue to play a role in human civilizations far into the future, as it stands.

7.3. MECHANICAL ENGINEERING

The design, manufacture, and maintenance of mechanical systems is called mechanical engineering. It's another very broad and diverse field that offers a lot of career options, if you're thinking about going into the field of mechanical engineering. You'll work with a wide range of mechanical processes and products, ranging from small component designs to extremely large plants, machinery or vehicles, as a mechanical engineer.

Some of the specialisms are completely different engineering fields in their own right and there are wide range of specialisms within mechanical engineering. As you consider your mechanical engineering options and interests, it's worth bearing this in mind.

As new technologies have emerged, the mechanical engineering discipline, which has evolved over the years, is one of the broadest engineering disciplines. Involving in the design, development, testing, and manufacturing of machines, robots, tools, power generating equipment such as steam and gas turbines, heating, cooling, and refrigerating equipment, and internal combustion engines are all done by Mechanical engineers.

Thermal/fluid systems and structural /solid systems are the major branches of mechanical engineering. As more efficient machines and power generating equipment and alternative energy-producing devices are needed the job outlook for mechanical engineers is also good. You will find mechanical engineers working for the federal government, consulting firms, various manufacturing sectors, the automotive industry, and other transportation companies.

Mechanical engineering is the study of objects and systems in motion, one of the most diverse and versatile engineering fields. The field of mechanical engineering touches virtually every aspect of modern life, including a highly complex machine, the human body (Figure 7.6).

Figure 7.6. Mechanical lab.

Source: Image by Wikimedia Commons.

A variety of career options beyond the industries listed above are the breadth of the mechanical engineering discipline allows students. A mechanical engineering education empowers students with creative thinking skills to design an exciting product or system, regardless of the particular path they envision for themselves; analytical tools to achieve their design goals; the teamwork needed to design, market, and produce a system and the ability to overcome all constraints. In many other fields, such as medicine, law, consulting, management, banking, and finance, these valuable skills can be applied to launch careers.

7.3.1. Application of Mechanical Engineering

7.3.1.1. Aerospace

Aerospace engineers are all about flight, whether that's planes, missiles or rockets. Designs that are more fuel-efficient aircraft that cut emissions, build the fleets of satellites that power modern GPS technology, and create the next generation of spacecraft for missions to Mars and beyond are all done by aerospace engineers.

In the manufacture, design, and testing of devices and machines both mechanical and aerospace engineers share their interest. Their job is to create equipment that runs safely, effectively, and efficiently, they also test and investigate the equipment to make sure that it meets the necessary criteria and measures. Aerospace engineers challenge flight mechanics and aerodynamics in a way that mechanical engineers usually do not, however, there is an overlap in the engineering principles in the both disciplines.

If one aspires to become a mechanical engineer within the aerospace field then they will have to become used to thinking like this. Aerospace engineers may take part with the designing and creating of satellites and weapons and usually go on to work with the government or military opposed to a regular mechanical engineer. To possess strong mechanical and mathematical skills, it is essential for a mechanical engineer in the aerospace field.

To operate design software and run simulations, they also will need computer skills. To perform most of their duties in an office, it's common for these engineers. Project leaders may need to work overtime hours and they will also work full time.

In the aerospace field a mechanical engineer will be involved with the design, manufacture, and testing of aircraft and the products that the aerospace industry uses. They will design spacecraft structures. Things like making sure that a satellite is able to withstand the force of a rocket launch, and its solar cells will work properly in orbit, and that its rocket engine is efficient with fuel use to reach orbit successfully, engineers are involved in. Projects like mission systems, airframe structural analysis and design, or flight tests they are specialized in such.

7.3.1.2. Automotive

The automotive industry is driven by mechanical engineers. From 80-seater buses to single seat F1 cars, they design bodyshells, wheelsets, and combustion systems for every type of moving vehicle. To find better ways to keep people moving, it's not all traditional fuels either – automobile engineers work with solar panels, hydrogen cells and other technologies.

Mechanical engineers can also be in charge of evaluating designs created by automotive engineers to ensure efficiency, safety, and reliability of the vehicles being produced, they are inherently good at designing machines, its components, processes, and systems. For example, that they can pass safety standards and be viable for commercialization, they can refine automotive models and its components.

Mechanical engineering skills can also be applied, in the calculation of power needed for cars, dimensioning the cars, their mechanical parts, calculating the required cooling system for a specific automobile and conducting and studying crash absorbing devices. Steering systems design and basically everything that involves hydraulics, pneumatics, and an automobile energy-efficient aspect, are done by them. Mechanical engineers handle the calculations of the processes that take place inside of automobiles.

There has to be close coordination of all the engineering trades, the production of automobiles is a multi-billion-dollar business so that everything works properly together. To develop engines, frames, braking systems, gears, and other components to make them work properly, mechanical engineers and automotive engineers have to work closely together in the automotive field.

Mechanical engineers complete some of the roles and in some cases automotive engineers are:

In vehicle development organizing and devising tests to answer questions from clients, engineers, and consumers involved (Figure 7.7).

Figure 7.7. Mechanical engineers at an automotive manufacturing unit.

Source: Image by Flickr.

- In order to design tests focused on Work with commercial departments, negotiating costs of engineering/development.
- Including projected costs, managing all details of projects.
- Including innovative and conventional methods creating new test methods.

- In project management and problem solving, bringing new products to market and being involved.
- Within a bigger team and being independent, developing an individual specialism.
- To keep colleagues up-to-date on problems, progress, and new development, being integral to regular team meetings.
- Acknowledging the benefits of engineering development to connected departments in order to secure internal funding and market projects.
- Supervising engineers, designers or analyzing the competence and validity of new technologies, operating in collaborative teams across countries.

7.3.1.3. Biomedical

Mechanical engineers change lives, working in the biomedical industry. For injured and disabled people, they create better, more lifelike artificial limbs to improve quality of life. Pacemakers, artificial valves and even robotic surgical assistants are all the work of mechanical engineers.

Through cross-disciplinary activities that integrate the engineering sciences with biomedical sciences and clinical practice, biomedical engineering is a discipline that advances knowledge in engineering, biology, and medicine, and improves human health. In healthcare; develop new procedures using knowledge from many technical sources; or conduct research needed to solve clinical problems, biomedical engineers design instruments, devices, and software used.

In research and development or quality assurance they frequently work. Electrical circuits, software to run medical equipment, or computer simulations to test new drug therapies also designed by biomedical engineers. Such as knee and hip joints are the body parts designed by them in addition. To make the replacement body parts in some cases, they develop the materials needed. Rehabilitative exercise equipment also made by them.

In developing design plans for products related to their field mechanical and biomedical engineers both spend time. While biomedical engineers can concentrate on a number of specialization areas that may emphasize medical research or developing methods to diagnose and treat medical conditions, mechanical engineers concentrate more on product manufacturing and fixing technical issues.

Biomedical engineers perform a wider range of tasks that includes training medical staff, publishing information about their research and conclusions, and developing ways to improve medical treatments, while mechanical engineers primarily focus on developing design plans, testing products and ensuring the devices are manufactured effectively.

In the medical field, biomedical engineers may also be involved with designing products that are used. From materials to software to diagnostic equipment, the range of products varies. To research and development of medical applications, treatments or diagnostic technologies related to acute or chronic medical conditions, biomedical engineers (also sometimes known as medical engineers) apply engineering principles.

By research organizations, manufacturers, government agencies or major medical centers medical engineers are employed. In the biomedical field, a mechanical engineer can be called a medical engineer. In medicine, an engineer's responsibilities can include research, development, testing, and evaluation of medical devices, advisement on new biomedical purchases for medical centers and hospitals.

In medical devices, sports medicine, stem cell research, biomarkers, genomics, nanotechnology, and more medical engineers are specialized. Artificial organs and prosthetic limbs, computer assisted procedures and treatments, informatics, and medical imaging are included in medical engineering projects.

In the biomedical field, a mechanical engineer is usually the person that is interfacing with the clients and managing the projects. Trying to find solutions and managing client interactions, they also contribute on some level to the technical aspects of the company.

7.3.1.4. Construction and Building

To focus on the detail's major construction projects, depend on mechanical engineers. This could mean designing the heating, cooling, and ventilation systems for a 28-story hotel, making sure a new metro tunnel project incorporates other services to make the most of under-city space or choosing the best way to deliver mains gas to an entire housing estate.

For the projects they're working on, civil engineers and mechanical engineers create design plans. They may oversee work on their projects and also need to be able to determine the expected costs related to their project.

While mechanical engineers focus on mechanical products, civil engineers are responsible for building things like highways and tunnels (Figure 7.8).

Figure 7.8. Mechanical works in construction.

Source: Image by Wikimedia Commons.

Mechanical engineers may work on mechanical systems in buildings, such as power generators or they may work with smaller items, such as medical devices. In the civil field a mechanical engineer will generally fill a very similar role to a civil engineer. There will be different situations, where the mechanical engineer is coming straight out of college or has previous mechanical engineering experience, and this will determine what area the mechanical engineer will focus on.

Mostly on the materials side, there is a lot of overlap between mechanical and civil engineering. The analysis of structures and machines are not that different (stress/strain, displacement/velocity/acceleration, etc.) The fluids and heat transfer analysis are the main differences (density, compressability, viscosity, etc.) that mechanical engineers do. Just to a limited extent civil engineers do thermodynamics too.

For licensure and further education there are concerns to be prepared. You are only allowed to perform within your ability, ethically, as a PE. To determine your competence, it is your discretion; and a master's degree may be the way to go if you don't feel that competent. It's probably within a mechanical engineer's grasp, for most civil disciplines (transportation, geotechnical, construction, and environmental). Before acceptance into a Master's in Structural Engineering, steel design, reinforced concrete design, and structural analysis will likely be required.

7.3.1.5. IT Sector

For public use, mechanical and computer engineering both involve the design, development, and testing of tools. To identify and solve problems each of these professionals works, using systems, modeling, and mathematics. In science and mathematics, mechanical and computer engineering are each rooted, and both require the ability to communicate complex ideas.

To develop and test their prototypes, as well as to analyze and design subsystems mechanical engineers use computer science and technology. In the latest software tools, those with skills such as 3-D printing, can even eliminate the need for prototypes, from concept to final product more quickly moving a project.

To build the ever smaller and more capable computer chips upon which computer scientists rely mechanical engineers use nanotechnology. In the computer field a mechanical engineer would help develop the computer programs and processes that allow mechanical engineers to analyze important data and conduct analyzes.

In the computer field specifically, a mechanical engineer, would be involved in designing and testing robots, helping to set up and test networks, perform routine checks to ensure systems and hardware are stable and operating efficiently among many other responsibilities. In order to ensure the organization is taking advantage of the most current technology, it is also advised for mechanical engineers to stay up to date on industry trends and technologies.

In the computer field, the success of a mechanical engineer will be aided by the abilities that they possess as engineers, such as their motivation to be able to work with minimal supervision and still meet deadlines, strong communication skills, both written and verbal, and able to break down complex technical problems into simple terms, superior attention to detail in order to spot minute errors in code and able to devise innovative solutions to problems and excellent creative thinking skills.

7.3.1.6. Electrical Field

However the largest difference lies in the types of products they investigate, design, develop, and examine mechanical and electrical engineers work in environments that can be compared performing similar tasks. In the electrical field a mechanical engineer will have to deal with the manufacture of electrical equipment, communication systems, in addition to designing electrical systems in vehicles, and navigation systems.

The creation of many electrical systems that work in both aircraft and automobiles their duties involved. To create and install the electric equipment is the focal point of mechanical engineers in the electrical field is which is provided in broadcast and communication systems, such as portable music devices and GPS devices. Such as hardware engineers they have to work in collaboration with other engineers, additionally.

Making it a career constantly on the cutting edge, they help to design and manufacture electrical products of all shapes and sizes. Mechanical engineers in the electrical field frequently work with computers, due to the job's complexities. While working in the electrical field. on the contrary from a mechanical engineer's normal duties, they will generally focus on generation of power and supply.

7.4. ELECTRICAL ENGINEERING

A professional engineering discipline that deals with the study and application of electricity, electronics, and electromagnetism is called electrical engineering (EE) (sometimes referred to as electrical and electronic engineering). In the late nineteenth century, the field first became an identifiable occupation with the commercialization of the electrical power supply and electric telegraph. Sub-disciplines including those that deal with power, optoelectronics, digital electronics, analog electronics, computer science, artificial intelligence (AI), control systems, electronics, signal processing and telecommunications are the fields that covers a range of sub-disciplines.

The term EE may or may not encompass electronic engineering. Where a distinction is made, electronic engineering deals with the study of small-scale electronic systems including computers and integrated circuits (ICs) whereas EE is considered to deal with the problems associated with large-scale electrical systems such as power transmission and motor control.

Another way of looking at the distinction is that while electronics engineers are concerned with using electricity to transmit information, electrical engineers are usually concerned with using electricity to transmit energy (Figure 7.9).

Figure 7.9. An electrical engineer at work.

Source: Image by Pixabay.

The largest engineering discipline is electrical and electronic engineering. Design, develop, test, and supervise the manufacturing of electrical equipment, including lighting and wiring for buildings, cars, buses, trains, ships, and aircrafts; power generation and transmission equipment for utility companies; electric motors found in various products; control devices; and radar equipment are all done by the electrical engineers.

Power generation, power transmission and distribution, and controls are the major branches of EE. Electronic engineers work on designing, developing, testing, and supervising the production of electronic equipment, including computer hardware; computer network hardware; communication devices such as cellular phones, TV, and; measuring instruments as well as audio and video equipment.

Computer and communication electronics are the growing branches of electronic engineering. Because businesses and government need faster computers and better communication systems the job outlook for electrical and electronic engineers is good. In job growth for electrical and electronic engineers as well consumer electronic devices will play a significant role.

7.4.1. Application of Electrical Engineering

- **Medical:** To benefit from EE advancements and solutions is the medical industry, which Astrodyne TDI plays a major role, which is one of the most prominent industries. The medical equipment that doctors use in their practices, as medicine gradually begins to change and advance more and more each day. Due to the fact that engineers have designed and created concepts which help the medical practice thrive, since the 1950's there have been huge medical transformations.

To benefit from the aid of electric engineering, one of the most prominent industries is the field of medicine. Innovations are being made regularly with more intelligent, doctors, and medical professionals are able to work with stronger and more reliable medical equipment, helping their patients. There have been numerous contributions made by skilled electrical engineers to the field of medicine, since the 1950s. Helping revolutionize the field and providing patients with more extensive medical care options for their illnesses, many medical professionals believe that even further advancements can be made.

- **Robotics:** In medical electronic engineering involves robotic surgery, one of the most significant recent inventions. Robotic tools are useful because they offer precision, flexibility, and extreme control, in minimally invasive proccdures. To perform surgeries that would otherwise be needlessly complex or impossible altogether the result is that surgeons can use automated capabilities. Robotic surgical technology is will assist and enhance their work, not likely to replace human surgeons (Figure 7.10).

Figure 7.10. Electrical engineering application in robotics.

Source: Image by Wikimedia Commons.

- **Virtual and Augmented Reality (VR/AR):** In EE, the development of VR/AR is one of the most influential trends. VR/AR is helpful for providing convalescent patients with an immersive way to participate in rehabilitation exercises, in medicine. To become familiar with new procedures or see 3D representations of difficult-to-visualize human anatomy, VR/AR is also invaluable in training tools for medical students — students can engage with augmented or virtual scenarios.
- **Consumer:** In consumer EE, many recent innovations have to do with electric vehicle capabilities and wearable devices.
- **Wearable Devices:** Wireless technology has been making exciting advances, in wearable consumer devices. Devices help users monitor their health and athletic performance like smartwatches. Devices can be smaller and more convenient to use because they run on smaller, longer-lasting batteries like wireless technology — often Bluetooth Low Energy.

In industrial applications innovations in wearable devices also have lifesaving potential. If they get too close to high-voltage equipment, some wearable devices can vibrate to notify engineers and without requiring a smartphone for access, they can provide valuable data.

If someone approaches sensitive equipment without the right wearable device, the machine will not grant access, wearable devices also increasingly have the technology to facilitate authentication. Electrical engineers' safety on the job will increase by these innovations.

- **Electric Vehicles:** In popularity, electric vehicles have steadily been gaining, and because of their energy efficiency and reduced carbon emissions, they are almost certainly the vehicles of the future. In market valuation, Tesla, for instance, recently rose to $100 billion — shows no sign of slowing down anytime soon and it is the first publicly traded carmaker in the United States to do so (Figure 7.11).

Figure 7.11. Tesla electric cars manufacturing unit.

Source: Image by Flickr.

By the year 2030, industry experts predict that, from only about 1 million at the end of 2018 the number of electric vehicles on the road in the United States will have ballooned to 18.7 million, up. In electric vehicle technology, heavy investments mean consumers have seen — and can anticipate — the emergence of various innovative improvements, including more powerful, longer-lasting batteries; enhanced charging technology; solar-powered vehicles and genuinely functional autonomous driving. There's even the possibility of electric airplanes.

- **Wireless Charging:** Wireless charging is one area of technology that holds particular promise for expanding the electric vehicle market. Wireless charging will likely eventually become standard for electric vehicles and it has some current applications for personal devices like laptops, smartphones, and earbuds.

Without the hassle of plugging in the car, an electric car owner will be able to park on a charging spot. It will become easier and more cost-effective to build, wireless charging docks will also be smaller.

- **Industrial Field:** A few different innovative technologies are emerging as game-changers, in the industrial field. In EE for industry here is some of the latest technology:
- **Augmented Reality:** For industrial use, advances in augmented reality are taking place substantially — 65% of VR/AR companies report that only 37% are working on consumer products, while they are working on industrial applications. Because it allows companies to simulate dangerous industrial scenarios without putting their employees through the actual risks VR is useful in industrial facilities. It superimposes data on a real visual to give engineers and technicians real-time information about the industrial systems they're working with and helps them take more informed approaches to repairs and maintenance AR is useful.
- **Smart Grid:** Industrial and commercial consumers can sell their surplus and even generate their own power. In part with the advent of smart grids, this development has changed electrical delivery infrastructure.

Including in homes, offices, and industrial facilities, smart grids contain smart devices throughout their infrastructure. To analyze trends and make more informed, efficient, and cost-effective choices about their electricity use, these smart devices collect and supply data that allows industrial facilities.

The devices can detect outages at once and notify the personnel who can rectify them and they predict surges in usage and prepare for the higher demand. By facilitating a quick resolution of any issues, perhaps most importantly, the smart grid allows for communication between the power company, distributors, and end-users and helps boost efficiency and lower costs.

- **Graphene Supercapacitors:** Supercapacitors store energy and have higher capacitance values and can function somewhat like rechargeable batteries and lower voltage limits than traditional capacitors. In place of activated carbon in their electrodes, graphene supercapacitors are supercapacitors that use graphene.

A supercapacitor, which can offer the advantages of increased energy storage, often store almost as much energy as a lithium-ion battery. While also providing high energy storage capabilities and charging incredibly rapidly, supercapacitors allow for the power density of capacitors — they can deliver a lot of energy in quick bursts (Figure 7.12).

Figure 7.12. Graphene supercapacitor.

Source: Image by explainthatstuff.com.

Graphene supercapacitors are ideal for high-frequency applications, it also helps enhance supercapacitors, because it is exceptionally conductive, so graphene, whereas traditional supercapacitors are not. Graphene has applications in computer processing units (CPUs) and ICs where standard capacitor materials do not because it allows for structuring and scaling down.

To combine with carbon nanotubules to help connect the geometrically unique graphene structures abled by graphene supercapacitors into a comprehensive network. This combination might reduce costs and boost capacitance and performance.

- **Artificial Intelligence:** In an industrial setting, artificial intelligence (AI) can help make electrical engineers' jobs much easier. In engineering work, they allow for several significant improvements, including:
  - For more complex and capable equipment, constructing AI and machine learning platforms;
  - For data analysis crafting complicated algorithms;
  - Enhancing current code or developing new codes; and
  - Processing images.

In particular, AI image processing, opens substantial new doors for industrial applications in engineering. Because AI allows for more sophisticated algorithms, image processing with AI is easier — they detect structural irregularities in equipment and send feedback to alert facility managers to the necessity of repairs, thereby promote safety in the workplace.

7.5. COMPUTER SCIENCE AND IT ENGINEERING

A variety of topics that relates to computation, like analysis of algorithms, programming languages, program design, software, and computer hardware encompassed by computer Science Engineering (CSE).

In EE, mathematics, and linguistics, CSE has roots. In the recent days has emerged as a separate engineering field but in the past was taught as part of mathematics or engineering departments.

The discipline that embodies the science and technology of design, construction, implementation, and maintenance of software and hardware components of modern computing systems and computer-controlled equipment it is known as Computer Science. As a combination of both computer science (CS) and EE computer engineering has traditionally been viewed (Figure 7.13).

Figure 7.13. Professionals at work in an IT firm.

Source: Image by Piqsels.

It has evolved as a separate, although intimately related, discipline over the past three decades. To solve technical problems through the design of computing hardware, software, networks, and processes, computer engineering is solidly grounded in the theories and principles of computing, mathematics, science, and engineering and it applies these theories and principles.

The study of computation, automation, and information is CS. Such as algorithms, theory of computation, and information theory, to practical disciplines including the design and implementation of hardware and software, CS spans theoretical disciplines.

An area of academic research and distinct from computer programming is generally considered by CS. The heart of CS is algorithms and data structures. The theory of computation concerns abstract models of computation and general classes of problems that can be solved using them.

The means for secure communication and prevent security vulnerabilities studied by cryptography and computer security. The generation of images is studied by computer graphics and computational geometry. Instead of only software engineering or electronic engineering, computer engineers usually have training in electronic engineering, software design, and hardware-software integration.

From the design of individual microcontrollers, microprocessors, personal computers, and supercomputers, to circuit design, computer engineers are involved in many hardware and software aspects of computing. This field of engineering focuses on how they integrate into the larger picture also on how computer systems themselves work.

To the description of computational processes and database theory concerns the management of repositories of data, approached by the programming language theory. Software engineering focuses on the design and principles behind developing software and human–computer interaction investigates the interfaces through which humans and computers interact.

The principles and design behind complex systems investigated by the areas such as operating systems, networks, and embedded systems. Construction of computer components and computer-operated equipment is described by the computer architecture. To synthesize goal-orientated processes aimed by AI and machine learning, such as environmental adaptation, decision-making, problem-solving, planning, and learning found in humans and animals (Figure 7.14).

Figure 7.14. Programming and computer science engineering.

Source: Image by Hippopx.

Computer vision aims to understand and process image and video data within AI, to understand and process textual and linguistic data aimed while natural language processing. CS determines what can and cannot be automated the fundamental concern. As the highest distinction in CS recognized by the Turing Award.

7.6. APPLICATION OF COMPUTER SCIENCE AND IT ENGINEERING

7.6.1. Health Informatics

While CS is about computers and their complex computational systems, healthcare is about the distribution of medical care to a community, CS and healthcare are two strikingly different fields. By studying the use of CS in the healthcare industry, a relatively new field called health informatics integrates the two.

When doctors began researching how computer logic could help diagnose and treat medical disease, health informatics first rose in prominence at the start of World War II (WWII). When computer technology advanced substantially, that health informatics gradually became a reality, it wasn't until the 1950s. The use of health informatics is promoted by the countries across the globe are implementing policies that promote.

Health informatics saves time, money, and lives. In the diagnosis and treatments of patients, combining a doctor's expertise with a computer's algorithmic results can reduce errors. Between patients and doctors, health informatics facilitates communication and makes remote medical care possible. Utilities like electronic health records and digital apps make it easier for patients to monitor their health and receive accurate, reliable medical advice, finally, computer technology comes with better coordination and management (Figure 7.15).

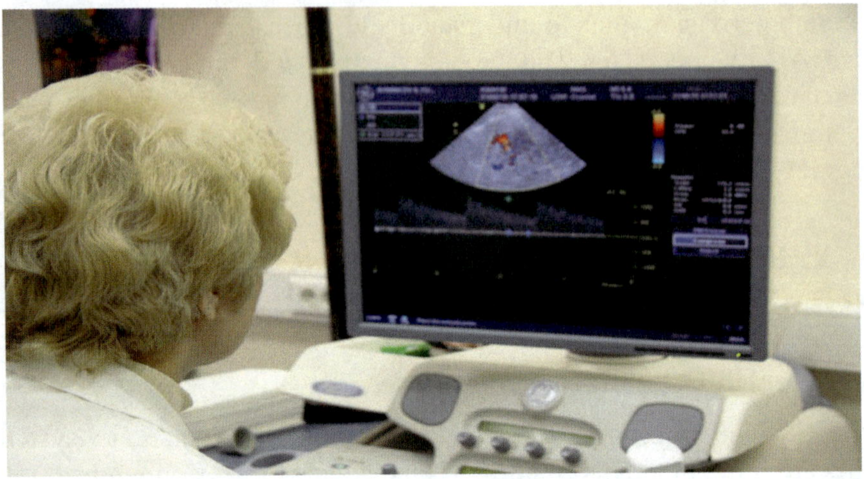

Figure 7.15. Computer science application in health informatics.

Sources: Image by Public domain picture.

Because health informatics is often divided into five subcategories, it is a broad field: translational informatics, clinical informatics, clinical research informatics, public health informatics, and consumer health informatics.

7.6.2. Consumer Health Informatics

From the perspective of the patient, consumer health informatics focuses on health informatics. To enhance the patient's experience with their healthcare, computer technologies are used in this subfield.

To help patients find medical providers, schedule appointments, and organize treatment plans web-portals used by hospitals, are developed by computer scientists who program and study these interfaces. For example, hospitals like the Houston Methodist relied on online platforms to allow patients to schedule coronavirus vaccine appointments online, during the recent COVID-19 pandemic.

These hospitals also utilized digital surveys to check in with patients remotely to ensure that no severe side effects were taking place, once patients received their doses. Increasing the need for computer scientists to design these virtual web-portals, today, roughly 4 in 10 individuals have filled out paperwork or made medical appointments online.

To help users monitor their health, mobile apps and wearable devices have also been programmed and designed by computer scientists. For instance, by Apple inc. the Apple Watch that was introduced in 2014, to track their heartbeat while resting and exercising incorporated a feature that allowed users.

This gave users the flexibility to track their heart rates without the need for a Holter monitor at their own convenience. Furthermore, newer versions of the same watch also detected irregular heart rates and notified users when this inconsistency occurred, in later years. It made it easier for users from the warnings, before health conditions significantly worsened to get medical help.

By using CS, medical records can also be digitized into electronic medical records or EMRs, which convert physical, complex medical records into a digital format, that reduced paperwork for patients. Accurate data about patients is being recorded and stored on a timely basis ensured by EMRs, which helps doctors diagnose and treat patients correctly.

Given the benefits of EMRs, to improve upon their healthcare systems, countries like India are working to incorporate the technology in their hospitals. To motivate physicians to use EMRs today, the U.S has even adopted federal policies.

Finally, to meet recent demands for telemedicine computer scientists are developing platforms like Zoom and Google Meet. In 2020, hospitals quickly reached their maximum capacities which resulted in more patients requiring remote care during the COVID-19 pandemic. Telemedicine is becoming a more convenient and necessary option for patients, as computers and phones become more widespread.

7.6.3. Public Health Informatics

From population data to public healthcare, public health informatics focuses on applying findings. To collect, analyze, and respond to large data sets obtained from populations computers are used. Primarily, to help users analyze current global health conditions, computers are used to monitor worldwide disease outbreaks in order.

To help users analyze current global health conditions, one example of this technology is Johns Hopkins University's COVID-19 map that was developed. When in the world coronavirus cases were increasing. Johns Hopkins successfully created a live visual that allowed users to stay up-to-date on information regarding the pandemic, by integrating data from different countries about coronavirus outbreaks with geographic data from satellites.

With the help of this users avoid certain areas of the world where cases were rising and educated the public on the global health status. To have computer scientists build programs that will efficiently, securely, and accurately collect data it is also important. For instance, by the addition of algorithms that output results in minutes, PCR, polymerase chain reaction, testing results used during the COVID-19 pandemic were enhanced.

Finally, to analyze population data and make decisions for the general public computer technologies are used. For instance, to analyze global weather patterns can notify weather reporters when a hurricane may be forming algorithms designed.

Authorities are better able to handle and respond to these adverse situations and protect communities around the world when computer programs give this signal. This task cannot be done as quickly, and the consequential delay could cost thousands of lives without CS.

7.6.4. Clinical Informatics

How CS can help deliver healthcare services studied in clinical informatics. In this sector of health informatics, modern technologies like AI play a significant role. Computer scientists make it easier for doctors to diagnose their patients with the correct treatments by designing software and programs that analyze specific data.

As an example, to help radiologists detect different forms of cancer MIT researchers used machine learning. It predicted which cancer disease was affecting the patient after the computer was given pictures of chest X-rays as well as other data. Implying that CS can make a positive difference in medical decision-making, the researchers found that pairing the results outputted by the computer with doctor expertise resulted in 8% more accuracy.

7.6.5. Clinical Research Informatics

To improve healthcare, clinical research informatics uses new data and findings. For studying, analyzing the results, and apply outcomes to the healthcare industry computer technologies are used.

Primarily, to perform clinical studies that give researchers new data to work with computers are used. As an example, to test a new drug called Alzhemed, results from large studies like the one performed by Neurochem Inc. to help patients fight Alzheimer's disease, are often collected, organized, and interpreted by computer programs, which was designed. This made the researchers to ensure that the drug tested works and can be reliably distributed to the rest of the population as it is tested.

For performing, CS also makes translational research easier. One of the largest fields undergoing translational study, today, CS plays a key role in oncology, by detecting mutations in DNA that can lead to insight on the cause of cancer.

7.6.6. Translational Bioinformatics

Finally, to store and analyze biomedical and genomic data, translational bioinformatics uses computers. With simulations, algorithms, and modeling, researchers can easily interpret complex data sets.

Computer algorithms, allow researchers to find patterns and analyze data as mentioned previously. As an instance, researchers often use Genbank, to interpret human DNA, an NIH genetic sequence database. Because the human genome contains over 3 billion pairs of nucleotides, which is too large of a dataset for humans to handle, computer programs are needed to find slight differences in genetic questions and identify these diseases.

In translational bioinformatics computer modeling is another technique used. Researchers can obtain information they can't easily see by modeling certain organs, sequences, or situations. At King's College in London, a group of researchers in order to infer certain properties like stiffness, which can potentially cause heart failures modeled a patient's heart. The researchers successfully analyzed the heart's stiffness and also gained new understandings of different mechanisms of the heart; the models created were incredibly beneficial.

7.6.7. Agriculture Field

In manual calculation, the application of the computer in agriculture research originally exploited for the conversion of statistical formula or complex model in digital farm for easy and accurate calculation which are found relatively tedious.

The same computers have been used to mechanization, automation, and to develop decision support system for taking strategic decision on the agricultural production and protection research in the next generation. Especially in the field of yield prediction, suitability of soil for particular crop, and site-specific resource allocation of agriculture inputs, etc. recently remote sensing and geographic information system (GIS) has placed a major and crucial role in agriculture research.

7.6.8. Data Mining

The process of discovering potentially useful, interesting, and previously unknown patterns is called data mining. To extract the metal, the process is similar to discovering ores buried deep underground and mining them. This process of converting data to information and then to knowledge described by using the term "knowledge discovery."

The data mining process requires an understanding of the decisionmaker's intentions and objectives, the nature and scope of the application, as well as the limitations of data mining methods and is interactive and iterative. People can focus on making the decisions, a variety of software systems are available today that will handle the technical details.

All most all statistical techniques we are using are just data mining, including bioinformatics either it may be in the field of agriculture, medicine or engineering.

7.6.9. Bioinformatics

In the areas of CS, information science and information technology, bioinformatics integrates the advances, to solve complex problems in Life and plant Sciences. Depended on chemistry, biology to make major strides, and this led to the development of biochemistry.

To explain biological phenomena at the atomic level led to biophysics is the need. The need to interpret the enormous amount of data gathered by biologists, requires tools that are in the realm of CS. To aid agriculture researchers in gathering and processing genomic data to study protein function is the role of bioinformatics.

7.6.10. Remote Sensing and Geographic Information System

The process of gathering information about an object, at a distance, without touching the object itself is the process of remote sensing. The photographic image of an object taken with a camera is the most common remote sensing method that comes to most people's minds. techniques have A unique capability of recording data in visible as well as invisible (i.e., ultraviolet, reflected infrared, thermal infrared and microwave etc.) part of electromagnetic spectrum are the techniques of remote sensing.

By human eye there are certain phenomenon, which cannot be seen, can be observed through remote sensing techniques i.e. the trees, which are affected by disease, or insect attack can be detected by remote sensing techniques much before human eyes see them.

A computer-based information system that can acquire spatial data from a variety of sources, change the data into useful formats, store the data, and retrieve and manipulate the data for analysis is called Geographical Information System. For private enterprises, government agencies, and academic institutions, today, GIS is a multi-billion-dollar industry and has become part of a basic information infrastructure (Figure 7.16).

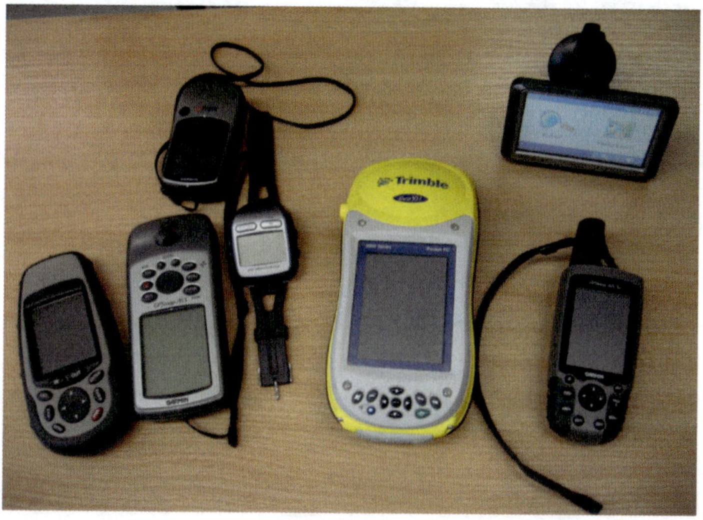

Figure 7.16. Computer science in making geographical information systems.

Source: Image by Wikimedia Commons.

For thematic mapping, handling spatial queries, and decision-making support, the majority of the operational GIS are used. In the field of agriculture and crop studies in India especially crop production forecasting covering both crop inventory and crop yield forecast models, soil mapping and soil degradation, drought assessment, command area monitoring, land suitability mapping, flood damage assessment, insect pest infestation forecasting and widespread availability of satellite signals that allow private use of GPS made it possible for farmers to spatially locate data from precision farming applications is the application of remote sensing data taken momentum.

To create resource database and to arrive at appropriate solutions/ strategies for sustainable development of agricultural resources, GIS technology is being increasingly employed by agriculture researchers. Due to improvement in space borne remote sensing satellites in terms of spatial, spectral, temporal, and radiometric resolutions the application of remote sensing and GIS techniques in the management of agricultural resources are increasing rapidly.

Buffer zones, neighborhood characterization, and connectivity measurement are the other analytical functions of GIS included. The ability to calculate more realistic distance measures among objects based on actual geometry, travel time, and cost, rather than straight-line distance is a particular feature of GIS.

To better integrate GIS with other software in statistical analysis, operations research, and AI tools is the trend in developing GIS analytical functions. Geographical information systems will likely become more user-friendly, looking into the future. There is an evolution from single-user systems to more open, multiple-user systems, with the development of web-based GIS.

7.6.11. Precision Agriculture

A system approach to re-organize the total system of agriculture towards a low-input, high-efficiency, sustainable agriculture Precision Agriculture is conceptualized. From the emergence and convergence of several technologies, including the Global Positioning System (GPS), GIS, miniaturized computer components, automatic control, in-field, and remote sensing, mobile computing, advanced information processing, and telecommunications benefits from this new approach.

In both space and time agricultural research is now capable of gathering more comprehensive data on production variability. The goal of Precision Agriculture is the desire to respond to such variability on a fine-scale.

7.6.12. Expert Systems

A specific kind of information system in which computer software serves the same function expected of an expert is called an expert system. As to the best choice of action for a particular problem situation. The hope is that we can design computers (information systems) that extend our ability to think, learn, and act as an expert, the computer, programmed to mimic the thought processes of experts, provides the decision-maker with suggestions.

The knowledge of experts without requiring their presence expert systems allow users to influence. In agriculture where experts are rare, expensive, or inaccessible expert systems are useful in any field especially.

The core component of any expert system since it contains the knowledge acquired from an expert in the field and from published literature is the knowledge base.

To build the knowledge base for the system typically, a knowledge engineer is responsible for working with an expert. A detailed analysis of the inference process and develop the prototype knowledge base performed by the knowledge engineer. In developing any knowledge base include knowledge acquisition, knowledge representation, knowledge programming, and knowledge refinement involved the tasks.

7.6.13. Decision Support Systems

To analyze complex information and help to make decisions are called decision support systems (DSSs) computer systems that provide users with support. With a specific function to help people with the problem-solving decision support systems are information systems – to some extent and decision-making process. A collection of people, procedures, software, and databases with a purpose consisted by DSS. In such systems the computer is the primary technology.

An advancement of management information systems, generally help human beings solve complex problems, and provide data that can lead to non-predetermined solutions that are beyond the limitations of expert systems are decision support systems.

Decision support systems may work actively or in a passive mode. Active systems give advice in certain situations, such as alerting technical personnel when a parameter being monitored exceeds its designated threshold value while passive systems are mostly used by decision makers or supervisors or physicians for reference purposes.

7.7. CONCLUSION

On all of humanity's activities, technology has a profound impact. Including the theory of the origin of the universe, the theory of evolution, and the discovery of genes, have given humanity many hints relating to human existence from civilized and cultural points of view by the technology inventions and discoveries. On the formation of our understanding of the world, our view of society, and our outlook on nature science and technology have had an immeasurable influence.

To the building and development of the civilizations of each age, stimulated economic growth, raised people's standards of living, encouraged cultural development, and had a tremendous impact on religion, thought, and many other human activities has led by the wide variety of technologies discoveries produced by humanity has led.

On modern society the impact of technology is broad and wide-ranging, influencing such areas as diplomacy, politics, the economy, defense, transportation, medicine, social capital improvement, agriculture, and many more. The fruits of technology fill every corner of our lives.

Among other expressions the hundred years of the twentieth century have been called the "century of science and technology," the "century of war," and the "century of human prosperity." Thus, technology have far brought humanity immeasurable benefits. In the twenty-first century, dubbed the "century of knowledge" and the time of a "knowledge-based society."

REFERENCES

1. Imeche.org. (n.d). *Where Do Mechanical Engineers Work* – IMechE. [online] Available at: https://www.imeche.org/careers-education/careers-information/what-is-mechanical-engineering/where-do-mechanical-engineers-work (accessed on 1 July 2022).
2. IT Hare on Software. (n.d). *Fields of Engineering Overview* – IT Hare on Soft.ware. [online] Available at: http://ithare.com/fields-of-engineering-overview/ (accessed on 1 July 2022).
3. Leach, J., (n.d). *The Different Types of Engineering (and Their Career Paths) Explained*. [e-Book] https://www.borntoengineer.com/. Available at: https://bvgs.co.uk/wp-content/uploads/2020/07/E16-Different-Types-of-Engineering.pdf (accessed on 1 July 2022).
4. Mechanical Engineering HQ, (2020). *Mechanical Engineer in the Biomedical Field* – Mechanical Engineering HQ. [online] Available at: https://mechanicalengineeringhq.com/mechanical-engineer-in-the-biomedical-field/ (accessed on 1 July 2022).
5. (n.d). *Computer Engineering* – ACM CCECC. [online] Ccecc. acm.org. Available at: http://ccecc.acm.org/guidance/computer-engineering#:~:text=Applications%20include%20consumer%20electronics%20(CD,pagers%2C%20personal%20digital%20assistants). (accessed on 1 July 2022).
6. Sharma, A., (2021). *The Importance of Computer Science in the Healthcare Industry by Angelica Sharma*. [online] County Line. Available at: https://crhscountyline.com/features/2021/10/26/computer-science-in-health/ (accessed on 1 July 2022).

INDEX

Printed in the United States
by Baker & Taylor Publisher Services